Instructor's Manual to Accompany

Presenting SERVICE

THE ULTIMATE GUIDE FOR THE FOODSERVICE PROFESSIONAL

SECOND EDITION

LENDAL H. KOTSCHEVAR, PhD, FMP
VALENTINO LUCIANI, CHE

PREPARED BY DIANE WITHROW

BICENTENNIAL 1807 WILEY 2007

JOHN WILEY & SONS, INC.

This book is printed on acid-free paper. ♾

Published by John Wiley & Sons, Inc., Hoboken, New Jersey.
Published simultaneously in Canada.

Wiley Bicentennial Logo: Richard J. Pacifico.

For general information on our other products and services, or technical support, please contact our Customer Care Department within the United States at 800-762-2974, outside the United States at 317-572-3993 or fax 317-572-4002.

Wiley also publishes its books in a variety of electronic formats. Some content that appears in print may not be available in electronic books. For more information about Wiley products, visit our Web site at http://*www.wiley.com*.

Library of Congress Cataloging-in-Publication Data:

ISBN: 978-0-470-23949-0

Printed in the United States of America

10 9 8 7 6 5 4 3 2 1

CONTENTS

Preface

Thank you for choosing to use ***Presenting Service, Second Edition,*** for your dining room service class. This Instructor's Manual contains the following resources:

- Each chapter states the **Learning Objectives** and contains an **Outline** of the chapter with bulleted notes for the classroom.
- Discussion and contemplation topics such as questions to pose to students and in class activities are interspersed, as appropriate, throughout the course outline for each chapter to help you to further engage your students in and out of the classroom setting.
- Each chapter contains a **Chapter Review Questions Answer Key** and **Chapter Cases Answer Key**, which provides suggested answers to the questions and case questions that appear in the text.
- The **Test** includes 5 True/False, 10 Multiple-Choice, and 5 Short Answer Questions for each chapter and a correlating Answer Key.
- Suggested **Activity** or **Training** activities can be found at the end of each chapter. These are topics that can be used in homework assignments or as part of a dining room lab situation

Chapter 1: A Historical Overview of Service

Objectives

- Describe the importance of excellent service to a successful operation
- Provide a historical overview of service
- Explain how haute cuisine developed, and how it influenced modern service

Outline

I Introduction

- The National Restaurant Association reports that nearly 50% of all customer complaints involve service as compared to 12% food, 11% atmosphere and 28% other

- Discuss last or most memorable complaint and categorize. See if class ratios hold true to NRA reports. **Variable- probably close though students may have a higher percentage for food.**

- Americans selected restaurants according to: (15 years ago/2005)

1. Quality of Service	Cleanliness (Sanitation)
2. Quality of Food	Quality of Service
3. Ambiance	Quality of Food
4. Price-Value Relationship	Price-Value

- Discuss the shift in priorities and speculate on reasoning. **Sanitation awareness is much greater today than it has been historically. Guests rely on posted health score to assist them in choosing whether or not to patronize a facility. Overall, ambiance may have improved to the point that it not as large of an issue**

II The Age of Service

- The economies of the world's most advanced nations are no longer based on the production of goods but for service industries

- This shift has happened due to increased productivity, disposable income and leisure time

- The Foodservice industry accounts for nearly 4% of the Gross Domestic Product and employs over 12 million people

- In 2003, sales were over $437M at over 900,000 eating and drinking places in the United States.

- Nearly one half of all adults living in the US eat out at least once per day.

- The foodservice industry is growing at probably a greater rate than many other service sector industries. Jobs are plentiful.

- Discuss the optimistic outlook for the industry. Ask students what their goals are and why they are taking the class**. Students with aspirations in the culinary field sometimes see little need to learn the front of the house. There are other students who may feel fearful if the class includes a student run dining room experience. All should be reminded that without the critical element of service, any restaurant can fail.**
- Foodservice is thriving and competitive. Difference is often excellent service in rising above competition.
- It costs very little more to provide good service than poor service.
- Discuss the costs of providing good service. How many of the earlier blunders discussed might be traced back to being understaffed? Poorly trained? Bad policies? Bad hiring? Which of these is associated with costs? **Having more people on the clock would appear to be more expensive unless it is considered that poor service costs customers and sometimes results in costly errors. Training is not free either but also can save more than it costs. Policies and hiring might be the least costly and can also save if done right.**

III Service: A Total Concept

A. The Meaning of Service

- The act of providing customers with a wide range of meal-related benefits and experiences. It is what makes people feel good about spending money to eat out.
- Service is not menial; it is a profession as servers are qualified persons of one specific occupation or field serving the needs of others.
- Physicians, religious leaders and teachers are all service providers.
- Professional people are supposed to meet the highest standard of moral and ethical behavior and treat others in a professional manner.
- Pride and self-esteem in the job can be had by mastering the art of service.
- Does the concept of being in a position of servitude carry negative undertones in a country where many have fought to overcome the stereotypes of race and gender? Is serving generally regarded as respected in the US? What can management do to assist servers in the development of pride and self-esteem? **While U.S. consumers want to receive service, they may be more reluctant to provide it for any number of reasons. Managers can oversee an operation that promotes professionalism and step in if guests behave in a way that denigrates servers. Managers can also discourage bad behavior by selecting dignified uniforms and treating employees with respect. Providing training to further professionalism both initially and ongoing must be part of any operation.**

1. The Tradition of Hospitality

    - The following two ancient rules of hospitality are customs of nearly every culture:

1. Guests should be made to feel welcome and wanted
2. All efforts should be made to see that no guests come to any harm. This encompasses the early Christian concept of *sanctuary*

- Modern laws holding hospitality operations especially liable for the safety of guests evolved from this old European value
- Why would the hospitality industry be more liable than other businesses? **People who are traveling and hungry are at the mercy of providers of those services, more so than businesses where patronage may be more voluntary. As hospitality has been seen as a religious duty, it is not surprising that the failure to perform may carry stricter penalties.**

2. Meeting and Exceeding Guest Expectations

- Service is often the single most influential factor in customers' decision as to where to eat out. It provides a competitive edge and keeps guests coming back
- Good servers do three things well:
    1. Pay close attention to detail
        - Results in good tips and recognition
    2. Work efficiently
        - Learn to plan and organize by:
            - Set up work stations carefully and replace supplies before they run out
            - Never walk empty handed if there is something to carry. Combine trips and save steps.
            - Stay organized and follow an efficient routine. Prepare for busy times and stay on top during slower times
    3. Are consistent despite challenges threatening demeanor
        - Focus completely on customers, showing a sense of urgency
        - Acknowledge every guest they have contact with
        - Good servers anticipate needs by watching and listening for clues, doing whatever is reasonably possible to please and thinking creatively
        - Ask class to respond to examples of what needs someone may have that could be anticipated- **Fries- catsup, Soup- bread or crackers, Beverage- straw and napkin etc.**

IV A Historical Overview of Service

- Early history can be gleaned from literary references and notes from events or practices of the people are sources of information

A. Beginnings

1. Ancient Times

    - Early people ate largely for survival with little ceremony. The discovery of fire allowed food to be cooked using crude clay dishware and utensils.
    - How do students suppose some of these discoveries were made? **Accidentally, for example, a lightening strike causing fire led people to experiment with what it could be used for. Clay that got left in the fire changed form usefully etc. Necessity also would cause people to be inventive.**
    - Improvements included perfection of design, development of pieces to eat and drink with, color and glazes and utensils such as ladles and cooking spoons.
    - After humans moved from caves to dwellings, fireplace cooking developed.
    - Diggings from modern Pakistan and ancient China showed existence of restaurant type units where the public went to dine.
    - Chinese served in fine pottery and porcelain dinnerware
    - Chopsticks were used since 6000 BC with knife, fork and spoon developed six or seven thousand years later in southwest Asia

2. Greek and Roman Times

    - *Taberna vinarias* were small eating and drinking establishments which *tavern* comes from
    - Wine was dispensed from huge stone jars preserved by pouring oil over the top
    - Cooked food was kept warm in *Thermopoliums*
    - Pagan feasts honored the gods with vast amounts of food and drink with celebrations lasting for days. Greeks feasted to Dionysus, the god of food and wine.
    - Ancient Romans rivaled Greeks in public feasts both by public officials and private citizens. Service was elaborate with four coursed banquets.
    - Many rare and exotic foods were served with beautiful dishes
    - Apicius, thought to be author of the first cookbook bankrupted himself with a banquet and committed suicide.

- Was dining one of the first forms of conspicuous consumption? **Probably one of the first ways to share or flaunt something universally desirable.**

3. Street Vending

    - Ancient Egypt and China had vendors selling foods in markets or at night, crying out the menu. This kind of vending continues in many places
    - Why might most of these sorts of vendors have been replaced by vending machines in the U.S.? **Sanitation concerns about food storage at proper temperature, quick service and delivery options make food more readily available to consumer.**

4. Service in Inns and Taverns

    - Inns and taverns were established to care for land and sea travelers.
    - Laws regulating inns including a Chinese provision that monasteries care for travelers
    - Discuss having monasteries care for travelers as a logical extension of the concept of sanctuary and hospitality as a religious duty. **Most world religions have stories about the need to feed the hungry and shelter the homeless. Some hospitality programs have Service learning requirements that may include volunteering at a homeless shelter. Projects of this sort might be well suited to a class on Service.**
    - These units offered limited service including for animals
    - Food was usually served but this was sometimes left to travelers or their servants to provide

B. The Dignification of Dining

- Ancient people began to attach philosophic or symbolic meaning to the act of eating so that they were expressing a feeling or belief
- These practices greatly influenced the kind of food served, how it was prepared and served. Very few cultures did not develop practices in these areas.

    1. Well Being

        - Certain foods were eaten primarily for health or sanitary reasons
        - People learned by trial and error, some cultures doing very well in achieving a diet that led to better health
        - The Chinese believed food was medicine and medicine was food
        - Many believed then and now that the act of eating certain food can bring about desirable results
        - A number of cultures used food and drink to signify delicate social relationships such as respect, love and other feelings

- How does our culture signify social relationships with food? **Jim Crow laws which barred African Americans from dining or drinking with those of European descent, sharing a meal with a friend or potential romantic interest through dating, family gatherings involving food, etc.**

2. Religion

    - Jewish people probably developed the most complex set of practices using food through the dietary or kosher laws in the Torah.
        - Only mammals with split hooves and chew cud are allowed
        - Mammals and birds must be slaughtered in a special manner and koshered or soaked and salted to remove all blood
        - Shellfish and pork are not allowed nor can milk and meat be eaten together or at the same meal
        - Dining is completely regulated and cooking is not allowed on the Sabbath
        - Seder dinner is on the first two evenings of Passover and symbolizes the escape of Jewish tribes from Egypt
    - Koran describes Muslim dietary laws allowing only *halal* food, which is Arabic for permitted
        - Pork, carnivorous animals, birds of prey, animals without external ears, most reptiles, blood and alcohol and intoxicants are prohibited.
        - Slaughter methods are also prescribed as well as a month of fasting from sunrise to set, Ramadan
        - Rules regarding the use and washing of hands are established
    - Buddhist philosophy also has established dietary practices
        - Eating meat is forbidden at certain times
        - Food is brought to the temple on feast days and laid at the feet of the Buddha statue
    - Hindu methodology relates how Prajapati created ghee (clarified butter) which is still considered precious
    - Bali rice farmers interweave the growing of rice into their religion
    - Christian religion uses food and drink for communion to symbolize the body and blood of Christ. Easter and Christmas involve special food and drink

- Do rules of other cultures seem far fetched? What about rules of the students own culture? **Students might be reminded that the Catholic faith formerly prohibited eating meat on Fridays. When the ban was lifted, many were simply not able to**

feel comfortable breaking the taboo and years later, are still following the old code. Ask if there are any other examples familiar to class members.

3. Socializing and Recreation

- Food and beverages supported public entertainment and social affairs
- *Colloquium* is Greek for gather together and eat
- Many other cultures used food and beverages to promote social life and entertainment
- Without electricity, radios, television, printed matter or transportation, food was the major concern and it is natural that it was used to enrich and extend peoples lives.
- Food of various classes was often similar, however tableware differed.
- Some cultures separated genders during dining, in others, they ate together. The separation of gender still exists in some Asian cultures
- In almost every culture, food and beverages were used as a means of worship or reverence
- Pagan cultures sometimes sacrificed animals or specially prepared and served them
- How does knowledge of food rituals affect the concept of service? **Many people have rituals with food that may be based on cultural custom or private practices. Many standards of service have become nearly ritualized so that guests have strong expectations about what is proper and right. The wine ritual of tasting etc. is an example.**

C. The Development of European Haute Cuisine

- From the 6th to 14th century, dining and culture progressed little
- Inns continued to serve travelers, offering three levels of service based on ability to pay
- Public life revolved around the church and monasteries continued to take in travelers on religious pilgrimages.
- The 15th century Renaissance brought art, music and architecture and fostered an environment where dining and service became more elaborate and sophisticated
- Guilds regulated the production and sale of goods including guilds involving food professionals whom effectively restricted their market
- In the 15th and 16th centuries, more ascended from poverty and demand rose
- Books on social etiquette appeared as a guide to desirable conduct in society including dining
- *The Courtier* and *Il Galateo* formed the foundation of hospitality service

- What is the point of etiquette? **Good manners seek to provide a means for people to feel comfortable in social situations and are thus connected with hospitality.**

  1. The Gastronomic Influence of Catherine de Medici

     - In 1533 King Henry II, the future king of France, married Catherine de Medici from Florence, Italy.
     - Catherine was appalled at the inferior level of food preparation and service
     - Diners helped themselves from common platters with their hands and ate from wooden trenchers with daggers as their only eating utensils
     - Catherine brought a staff of master cooks and servers, linens, fine dishes and goblets
     - She introduced society to forks, knives and spoons
     - Delicious food was now served in a lavish manner
     - When Henry IV ascended the throne, he followed his aunt's lead and France's nobility became connoisseurs of fine food, drink and service
     - Those who were not nobility or privileged continued to eat and drink simply in their homes, inns and taverns
     - Is it surprising that the French had been so behind when their reputation as gourmets is stereotypical? **Students are often surprised and interested to learn this information**

  2. The *Restorante*

     - In 1765 in Paris, Boulanger opened the first restorante claiming his soups and breads were healthful and could restore energy, hence, *restorante*
     - Guilds objected and sued him for infringement of their right to prepare and serve food to the public exclusively
     - Boulanger countersued, and with the help of friends in high places, won opening the door for others to start similar operations
     - The foodservice industry, as we know it, began
     - It seems foodservice was around since fire was discovered. Why do these events have such significance for our industry? **It marks the beginning of restaurants as an entrepreneurial venture that can be embarked on by anyone who desires to pursue it. It also introduced the word associated with our modern commercial industry.**

- Coffee was introduced to Europe in the 17th century bringing about the development of the coffeehouse which served other beverages and sometimes light food
- The French Revolution ended the rule of the kings and a new middle class arose and begin patronizing restaurants, increasing the market for high quality food, drink and service
- Cooks and servers who formerly served the upper class found jobs in the new industry.

3. Discriminating Gourmets
   - As the middle class grew, discriminating gourmets begin to write about eh art of fine dining including books on taste, a gourmet magazine, and dictionary
   - Simultaneously, Marie-Antoine Carême trained and led a group of chefs to continue his high level of food service. They invented new dishes, service and established rules on food combinations and order
   - What did all that enthusiasm accomplish for the common person? **As standards rise for the elite, so they rise for the humble. Love of food should be a universal concept.**

D. The Growth of Service in Modern Times

- Service development after 1900 revolved around the tremendous growth of the industry due to industrialization, mobility and disposable income.
- Today, 25% of all meals are consumed away from home representing 42% of the total food and drink dollars spent

1. The Rise of Hotels
   - Greater mobility led to hotels which affected food service
   - The joining of fine hotels, fine dining and fine service culminated in the partnership of Cesar Ritz and Auguste Escoffier a hotelier and one of history's greatest chefs
   - They made an unmatchable team and soon had many imitators
   - In 2005 more than 46,000 lodging properties with 3.8 million rooms have been built
2. Restaurants and Service in the United States
   - The first taverns in the U.S. were patterned after those in England increasing in number and becoming an essential part of colonists lives
   - In 1656 Massachusetts passed a law requiring every town to have at least one tavern as they served as a meeting place for discussions and news

- The Exchange Buffet, a self-service cafeteria style operation was the first recorded restaurant in the United States built in the early 1800s opposite the NY stock exchange
- Other historic restaurants include; Delmonico's, Sans Souci and Niblos Garden. Boston's Union Oyster House, opened in 1826 is still in operation
- Around 1900 the U.S. industrialized and workers began eating away from home more often.
- City dwellers, institutional food service and mandated public school lunches made dining out a common experience
- Today the foodservice industry is one of the countries largest numbering 900,000 eating places doing $437 billion in sales employing more people than any other industry including servers
- What new things have you learned that interested you? What trends today might be traced to historic roots? **Various**

Chapter Review Questions Answer Key

1. Churches, monasteries and convents were recognized as places of protection, referred to as *sanctuary*. This concept strongly influenced how guests were treated.
2. Pagan feasts were held to honor the gods of ancient times. Greek feasts lasted for several days. Their feast to Dionysus, the god of wine, was characterized by merry making.
3. Dietary laws and customs of various world religions such as the Jewish faith, Muslim, Buddhist, Hindu, and Christianity all dictate or suggest how, when and what food is consumed and what food may be taboo. Holidays have symbolic meaning with food used to make offerings or in memory or honor of events which inspired the holidays. For example, Jewish kosher laws dictate that pork is forbidden. Accepted meat must be slaughtered in a particular fashion. Cooking on the Sabbath is forbidden. Koran also prohibits eating pork as well as carnivorous animals, birds of prey, earless land animal and most reptiles, alcohol and intoxicants. Laws dictate the manner of slaughter of approved animals. Buddhist religion sometimes prohibits meat and proscribes the manner of consumption of approved foods. Food is laid at the feet of Buddha in the temple on holidays. Hindu myths relate how the Lord created clarified butter which is still considered precious. Communion is an example of ritualistic use of food in Christianity.
4. Catherine de Medici moved from Italy to France in 1533 to marry the future king of France, Henry II. She was shocked at the inferior level of food preparation and service. Diners ate from large common platters, helping themselves with their hands, filling wooden trenchers and using only daggers as flatware. Bones and waste were cast to the floor for household pets. Catherine brought a staff of cooks and servers, tablecloths and napkins, fine dishes, carved goblets and flatware. Food was now delicious and dining for the noble class became refined.

5. Guilds were food professionals such as the *Chaine de Rotessiers* (roasters of meat), *Chaine de Traiteurs* (caterers) and *Chaine de Pattissiers* (pastry makers) who formed to help regulate the production and sale of their goods. Their power effectively restricted their market. They had the right to sue to stop those who they felt were infringing on their trade.
6. Boulanger opened the first *restorante* in Paris in 1765. His soups and breads were served to promote digestion and restore energy. Guilds sued him but he countersued and, supported by friends in high places, won the right to compete with the guilds. This opened the door for others to start similar *restorantes* and the food service industry began.
7. Marie-Antoine Carême trained a large number of very famous chefs to follow him and continue his high level of food and service. They invented new dishes and service, established rules on what foods should be served together, when they were to be served and the manner of service.
8. Industrialization contributed to the foodservice industry by introducing the worker to eating out. Cafeterias and lunch counters sprang up to serve blue-collar laborers and white collar professionals.
9. Guest safety and security from harm has evolved in modern times to innkeeper and restaurateurs liability for guest safety as been codified into law.
10. Service is often the single most influential factor in customers' decisions as to where to eat out. Great service provides a competitive edge and keeps customers coming back. Good servers pay attention to detail, they work efficiently and are consistent.

Chapter Cases Answer Key

Excavating Ancient Ruins

What pots and pans or cooking implements might provide clues as to what sort of food was prepared. Spits, for example, would indicate large pieces of meat. A study of the climate and indigenous plants and animals of that period would also indicate what foods were available. Fire pits or ovens as well as any pots and pans would help indicate methods of preparation. Dishes, flatware (or not) would help to see if food was served from a common platter and whether or not there were individual dishes for eating from. A flat dish might indicate solid pieces of food while a curved bowl, more liquid such as stews or soups. If bones were found on the floor, it may be an indication that leavings were simply tossed to the animals. It is unlikely that fabric would remain but that might be telling as well.

Ancient Foods

Various- Ancient people must surely have eaten what was at hand. Animals, fish, and birds which lived within spearing, trapping or netting range were surely hunted and killed for protein. When animals became domesticated, eggs, milk and breeding for slaughter would have simplified the hunting process. Plants that grew indigenously were surely also gathered to be eaten. It would be a trial and error process to discover what was and was not poison or palatable.

Chapter 1 Test

True/False

T F 1. The economies of the world's most advanced nations are based largely on the production of goods.

T F 2. Sanitary standards, followed by service are the number one and two factors in the selection of a restaurant by consumers.

T F 3. Though it is hard work, serving is not a profession.

T F 4. The Irish custom of offering guests a pinch of salt and small glass of wine sprang from a desire to preserve food, banish bacteria, and promote good health

T F 5. Street vending is a recent development in the world of food service.

Multiple Choice

6. The food service industry will
   a. Remain stable as a modest part of the economy
   b. Decline as the Cooking Channel and Food Network piques consumer interest in eating at home.
   c. Will continue to grow, probably at a greater rate than many other service industries
   d. Will rise and fall in terms of demand, depending on competition from groceries, service stations and vending.

7. The meaning of Service is:
   a. Taking orders, placing down food and clearing up after.
   b. The act of waiting on the customer
   c. A sometimes demeaning state of affairs whereby the server is at the beck and call of the sometimes unreasonable customer in order to receive pay in the form of gratuity. This pay may or may not occur as it is at the whim of the customer.
   d. The act of providing customers with a wide range of meal-related benefits and experiences.

8. Anticipation of needs refers to:
   a. The notion that the guest always wants *something*
   b. The appetite of the guest which allows for suggestive selling
   c. The knowledge that the kitchen wants the servers to pick up food from the and keep special requests from guests to a minimum.
   d. Watching and listening carefully for clues, doing whatever is reasonably possible to please and thinking creatively when serving customers

9. Which of the following was probably the FIRST to be developed in relation to historical food service?
   a. Glazed ceramic cookware
   b. The discovery of fire

c. Chopsticks
d. Knife, fork and spoon

10. The vessels used in ancient Greek times for keeping food warm were called:
    a. Thermopoliums
    b. Taberna Vinaria
    c. Mensa Primara
    d. Apicius

11. The customs which have evolved around Well-Being and food service involve:
    a. Health and social relationships
    b. The good feeling of being full after feeling hungry
    c. The enjoyment of sharing food with friends
    d. Having nutritional needs met by food that tastes good

12. The relationship between food and religion may be characterized as:
    a. The custom of many of praying before a meal
    b. Using food practices to symbolize belief systems through rules, customs and rituals
    c. little importance to our industry, particularly in institutional food service where church and state are restricted from interfering in the affairs of the other.
    d. Both represent ancient needs which humans have fought bloody battles over to obtain domination

13. What contribution did Catherine de Medici make to the foodservice industry?
    a. She introduced flaming desserts, Caesar Salad and other tableside preparation methods using the *gueridon.*
    b. She inspired the use of food to promote health and well-being through the development of an understanding of nutrition
    c. Shocked at the appalling state of French food and manners, she introduced fine food and flatware to the nobility.
    d. She invented the appetizer as a way of tiding over her lover, Count Florence, until the dinner hour in order to keep him in her chamber.

14. Partner to hotelier, Cesar Ritz, he adapted and simplified the elaborate classic menus of his time to highlight top-quality cuisine and service:
    a. Boulanger
    b. Marie-Antoine Carême
    c. Auguste Escoffier
    d. Prajapati

15. The first recorded restaurant in the United States was:
    a. The Exchange Buffet
    b. White Castle
    c. Boston's Union Oyster House
    d. Boulanger's

Short Answer/Essay

16. How does the number one factor in restaurant selection and the term *sanctuary* relate to one another?
17. How did foodservice establishments affect entertainment in historical times?
18. How did etiquette as codified by such texts as the Courtier, and Il Galateo relate to the hospitality industry?
19. How did gourmets such as Brillat-Savarin and Alexandre Dumas pere influence the food service of their time?
20. How does the foodservice industry fit into the U.S. economy?

Chapter 1 Test Answer Key

True/False

1. False – we are in the Age of Service
2. True – fifteen years ago, Service was followed by Food with sanitary standards not even listed as a factor.
3. False – servers are qualified persons in one specific occupation or field serving the needs of others just as attorneys, physicians and other professionals do.
4. False – the custom was developed as both were precious commodities and by offering them, the guest felt truly welcomed and wanted.
5. False – street vending developed early. Ancient Egyptian tombs show vendors selling food in markets. Street eating was also common in ancient China. This kind of vending is still done in many countries.

Multiple Choice

6. c
7. d
8. d.
9. b
10. a
11. a
12. b
13. c
14. c
15. a

Short Answer/Essay

16. Cleanliness and sanitation relate to the feeling of safety and welcome that the guest receives and to what makes a guest feel safe. *Sanctuary* is an ancient term that meant that a guest should come to no harm under the care of the hospitable host.

17. Before cable television and the motion picture industry, people got together in taverns, street stalls and other eating and drinking places and gathered socially. Dining and interacting with others was a major form of recreation.

18. As more people ascended from poverty during the 15th and 16th centuries, demand for better service and cuisine rose, particularly in Italy. Books on social dining and etiquette appeared including acceptable behavior while dining, dining room décor, and desirable conduct in society. These works formed the foundation of hospitality service.

19. Their books on the art of fine dining helped to fuel interest in the further development of high levels of cuisine and service. To satisfy the gourmet palate, new dishes and service and rules on what foods should be served together were developed by chefs such as Marie-Antoine Carême.

20. The foodservice industry is one of this country's largest industry numbering nearly one million eating places doing well over $400 billion in sales. It is among the top ten industries and employs more people than any other.

Activity/Training

- If this class is associated with a student dining room, this chapter marks the beginning of training. Students are shown the fundamentals of the layout of the facility, how to operate a point of sale or cash register system, order placing and tray handling according to the policies and procedures of the location. **Now is a good time to stress the concepts of attention to detail, working efficiently and consistently. The examples on page five should prove to be helpful.**

- Ask the class for some of their first service experiences and how their perceptions of service and good service have evolved. If early memories are centered on drive through and delivery dining, what elements of those sorts of service made any particular event stand out? If experiencing a children's menu at a family casual or fine dining establishment, what made the experience responsive to their needs at that time? What needs did their parents have? Do students continue to patronize any favorites from childhood? What are their current favorites and why? **Answers will vary with memorable experiences generally on the extreme end of the spectrum. Entertainment value may rate as high as or higher than food value for young children. Parents want their children to be content and relatively well behaved (quiet) in order not to disturb other diners.**

- Ask students to begin to develop their own criteria for what "good service" is when dining out. This might be formalized into a restaurant critique assignment using a self-developed form and rating system. **Different people's priorities and peeves will vary with some commonalities. The amount of dining out experience and sophistication will likely affect perceptions but not necessarily.**

Chapter 2: The Professional Server

Objectives

- Outline the skills and behaviors common to professional servers
- Explain how hiring qualified employees helps an operation deliver excellent customer service
- Describe laws that affect employees and employers

Outline

I Introduction

- This chapter focuses on finding work, professionalism and the laws affecting servers.

II Finding Work

- Finding a server job includes getting a good fit. This means knowing the operation. Word of mouth, help wanted ads, union headquarters, employment agencies and just walking in are all methods of locating a job.

A. The Interview

- Present yourself as a desirable employee
- Be positive, explain qualifications, be mindful of body language
- Dress and grooming should match tone of operation
- Managers should hold interviews only when seriously interested in candidate
- Interviews should be designed to gather information. Open ended questions are good for this
- Ask students about interviews that they have had or given. Practice answering some of the open ended questions that appear in exhibit 2.1

B. Interviewing

- Success may hinge on actively hiring and retaining the most qualified, talented and motivated employees.
- This process includes screening, effective interviewing and selection
- Interviews are divided into four parts:
    1. Preparation
    2. The Interview
    3. Ending the Interview
    4. Evaluation

- The key to conducting effective interviews is to plan in advance
- Job Description Specifications are written definitions of the requirements of the job and the person that should be hired to fill these requirements.
- Set up questions to get information needed to judge candidates suitability for the job covering: education, motivation, ability to work with others and relevant personal characteristics
- Ask students what they learned if they tried out exhibit 2.1 questions previously.
- Ask open ended questions and use most of your time listening carefully to the answers
- Interview room should be private, orderly and unintimidating – sit next to the person being interviewed, not behind a barrier creating desk or table in order to put candidate at ease.
- Make arrangements not to be interrupted
- Have a note pad ready to take notes
- The applicant is also evaluating their interest in working at your location. Be positive and encourage the candidate to speak. Do not expect spontaneous answers
- Do not mislead or make false promises
- Give a tour and introduce the candidate to others. Ask if they have any questions.
- In ending, thank the candidate and give them information on the process
- After the interview, review notes, summarize, weighing all the facts.
- The text does not mention checking references. Ask the class their views on this practice.

C. The Legal Side of Hiring

- Federal laws are covered in the text but local laws must be followed as well. Failure to observe employment laws can result in heavy fines and severe penalties.
- Only job related questions should be asked of the candidate. Careful planning results in questions that are legal and effective.
- Questions regarding the following are prohibited by law:
    - Race, religion, age or gender
    - Ethnic background
    - Country of origin
    - Former or maiden name and parents names
    - Marital status or information about spouse

- Children, child care arrangements, pregnancy or future plans to become pregnant
- Credit rating or other financial information, or ownership of cars or other property
- Health
- Membership in an organization
- Voter preference
- Weight, height, or any questions relating to appearance
- Languages spoken unless the ability to speak other languages is required in the position
- Prior arrests (it is legal to inquire about convictions)

- Ask class why above questions are prohibited and relate to laws in exhibit 2.2 or assign groups to match the illegal question with the law prohibiting it.

III Looking Professional

- A positive first impression can translate into higher tips, better shifts and sections and advanced opportunity for recognition and advancement

A. Uniforms

- Bars, family style and casual restaurants uniforms feature; khaki pants, shorts, t-shirts, cotton button-downs or polo shirts
- Servers in upscale fine-dining restaurants often wear; ties and black aprons over formal clothing
- Institutional food service uniforms typically resemble traditional uniforms
- All should be clean, wrinkle free, and well fitting. Shoes should be comfortable, sturdy and skid proof
- Pens, order pads corkscrews or matches should be in pockets below the waist to prevent them from dropping
- Extra uniforms should be on hand
- Jewelry should be limited due to sanitary reasons
- Personal hygiene is vital to food safety and sanitation
- It is preferable to change into the uniform at work. If impossible, servers should come directly to work in their uniform
- Hair must be restrained, nails kept short and clean
- Gloves or tongs should be required when handling food. Sores, cuts or infections should be covered by Band-Aids
- Nothing that touches the mouth of the guest should be touched by the hands of the server

- Discuss sanitation as it relates to uniforms. How does providing more than one uniform assist servers in staying clean? Discuss uniforms students have liked and disliked and the reasons.
- Discuss provocative uniforms as they relate to professionalism.

IV Demeanor and Attitude of Successful Servers

- A pleasant and positive attitude has as much to do with success as knowing how to do the job. A survey showed only 10% were fired for not knowing the job, the other 90% due to personal traits or negative attitudes
- Discuss students' experiences with coworkers or as customers in dealing with a server (or other employee) with a bad attitude. Can attitude be determined at the interview stage? **Various responses.**

A. Delivering what Customers Expect: Operation Knowledge

- Servers must know complete information about the company for which they work and the products offered

B. Maintaining a Positive Attitude

- One of the most important traits a good server must possess is a positive attitude toward all aspects of the job
- Good servers correct their mistakes and learn from them
- Setting small goals build confidence
- Make an earnest effort to be friendly to and to please guests
- Servers should never talk about personal problems or inappropriate topics. The guest should be the focus of attention.

C. Courtesy

- Courtesy means being polite, gracious and considerate toward others. It means putting the guest's needs before one's own.
- Even difficult guests may be won over with courtesy, if not the mistreatment should not be taken personally
- Discuss class members experience with a difficult customer and how it was handled.

D. Tact

- Tact is the art of saying and doing the right thing using the right words at the right time. It is also an intuitive sense of what to day and do
- Remembering names and using diplomacy demonstrate tact as does handling sensitive situations so that everyone involved is left with their dignity
- Discuss situations where tact is required. Role play can be effective here.

E. Sincerity and Honesty

- Natural, not forced or phony behavior demonstrate sincerity and honesty
- Admitting mistakes and correcting them are part of being frank and truthful

F. Camaraderie

- The ability to get along with people enhances professional advancement

V. Learning Skills

- Learning and training is a continuous process and includes DVDs, study courses, computer programs, CD-ROM programs, simulations, online courses, training sessions, and other servers

VI. Product Knowledge

- Servers should be able to answer all guest questions about the menu
- Servers learn the menu by studying it, knowing all specials, tasting, and learning ingredients and preparation methods.
- Servers should know; menu items, signature items, specials and promotions, estimated run out time and items of interest
- Ask students if any have serving experience and if so, how they learned the menu. Ask what methods would be most effective.

VII. Suggestive Selling

- Suggestive selling involves offering guests the full range of products and services available.
- Suggesting items results in increased tips, guests who enjoy their experience more and increased profit to the operation
- Knowing menu terminology is essential

VIII. Organization

- Disorganization breeds a frantic pace, tension and nervousness
- Good servers have a rhythm, anticipate guests needs and are observant

IX. Tips

- Tips means "To Insure Prompt Service"
- Servers should receive a fair tip for good service
- Servers who welcome and thank guests and give extra effort needed to please should far outdistance their peers in tip earnings

- Tips are considered part of a server's salary and must be reported to the Internal Revenue Service. The employer must report 8% of each servers total sales
- Discuss issues surrounding tips with the class. What other professions rely on customers to provide the bulk of their pay? **Strippers** In professions where commission makes up the bulk of the income, is that amount set by the customer or is prescribed? **Commissions are usually set by management and not subject to much negotiation.** Discuss the European practice of adding a service charge.

X. Unions

- Servers are represented to management by union representatives who charge their members dues
- Contracts cover job classifications, duties, scheduling, pay, grievances, vacation, work week, breaks, sick leave, termination etc.
- Written job specifications an descriptions are advisable to have prior to management signing the contract so that function is not done by the union
- Union contracts often require a process leading up to termination which includes oral and written warnings. These are good practices in any case.
- Management should try to settle employee matters internally but if grievance is not satisfied, the union can take it on if it involves a contract violation. If not settled, a grievance committees decision is binding

XI. Laws Affecting Servers

A. Privacy Act

- Forbids employers from asking non-job related questions that might discriminate against a group of qualified job applicants

B. Fair Labor Standards Act

- Protects workers between 40-70 from discharge because of age- it also covers teenage workers, hours and union activities

C. Family and Medical Leave Act

- Requires employers with 50 or more employees to offer up to 12 weeks of unpaid leave for the birth, adoption or foster care of child, care for an immediate family member with a serious health condition and care for self with serious condition.

D. Civil Rights Act

- Prohibits discrimination against job applicants and employees on the basis of race, color, religion, sex or national origin.
- Further it covers wrongful discharge on the basis of age, disability and participation in union activities

- Sexual harassment is now covered and defined as unwelcome sexual advances, requests for favors, and other verbal or physical conduct when:
    - Submission is made a term or condition of employment
    - Submission or rejection is used as a basis for employment decisions affecting that person or
    - Such conduct has the purpose or effect of unreasonably interfering with a person's work performance or creating an intimidating, hostile or offensive working environment
- Discuss the above laws with students and their experiences with them or views on them. It is usually easy to get a lively discussion going on these topics.

E. Beverage Alcohol

- Sale of alcohol is regulated by state and local laws covering licensing, permits and how liquor may be sole
- Control states handle the sale and distribution while others permit retail establishments to purchase from prescribed dealers
- Dram shop laws hold anyone serving alcohol liable if an intoxicated patron injures or kills a third person, thus the term third party liability.
- Damages can be large as can insurance premiums covering them
- In order to reduce liability, many jurisdictions require training in responsible service of alcohol
- Responsible service involves never serving to minors, intoxicated patrons or habitual alcoholics
- Verifying identification thoroughly through photo identification can prevent service to a minor
- It can be difficult to deal with intoxicated patrons. If they try to drive away the police should be called if attempts to stop them fail
- Discuss class experience with being carded, alcohol related accidents involving injuries etc. Discuss purpose of third party liability (profit in sale of dangerous substance involves responsible sale of that substance. Alcohol, weapons and explosives all charge vendor with particular care in their sale) Class discussion is generally lively

F. Immigration Reform and Control Act

- Makes it illegal to hire aliens not authorized work in the United States
- Employers must verify citizenship or work permit
- Every new employer must complete a Form-I-9 upon hiring to verify that the new employee has submitted satisfactory proof of identity and work authorization if required

- Discuss the historical reliance of our industry on immigrant labor and the political stance of the National Restaurant Association on immigration for the purpose of employment

G. Americans with Disabilities Act

- Prohibits discrimination against people with differing levels of ability and requires reasonable accommodation for employees with disabilities
- It requires public places and services to accommodate guests with disabilities
- Ask students about their experience working with persons with disabilities. **Remind students of the shortage of labor in our industry and the need to recruit and retain creatively.**

Chapter Review Questions Answer Key

1. A grievance committee is appealed to when a server has a grievance which is not satisfied by the workers own efforts followed by an attempt by the union to settle the matter amicably. The union contract states that both parties will be bound by the committee's decision.
2. It is not an appropriate question. The appropriate question would be to ask if the candidate is able to perform the functions of the job. The employer is prohibited from discriminating against qualified individuals and must make reasonable adjustments in facilities and practices to permit participation of disabled persons.
3. Servers can maximize tips through their actions and words. They can always welcome and thank guests and give extra effort to please them. Performing the technical skills of serving such as serving from the left and removing from the right and handling dishes, glass and flatware in a sanitary fashion and being observant to guest needs and responding proactively will contribute to tips.
4. Sexual harassment is defined as unwelcome sexual advances, requests for sexual favors and other verbal or physical conduct of a sexual nature when; submission is made a term or condition of employment, submission or rejection of such conduct is used as the basis of employment decisions or such conduct has the purpose or effect of unreasonably interfering with a person's work performance or creates an intimidating, hostile or offensive working environment.
5. Third party liability refers to dram shop laws that hold anyone serving alcohol liable if an intoxicated patron injures or kills a third person (hence the term). It relates to laws that require those who profit from the sale of alcohol to serve it responsibly.
6. A typical server's uniform for a casual operation consists of khaki pants, shorts, t-shirts, cotton button down or polo shirts. The uniform should be clean, wrinkle-free, and well fitting with comfortable sturdy shoes for standing, walking and contact with slippery surfaces.
7. Tact is the art of saying and doing the right thing, using the right words, at the right time. It is an intuitive sense of what to say or do in order to maintain good relations. When a guest's credit card has been denied, it means handling it in such a manner that the guest does not feel

embarrassed and is left with their dignity. Speaking quietly in order to keep the matter a private one, sympathizing and remaining nonjudgmental would help in this situation.

8. Job specifications list knowledge, skills and abilities, work experience and education and training that the person should possess in order to be able to perform the job.
9. Some ways to show that you are actively listening in an interview are by; nodding, maintaining eye contact, asking questions and, at times, repeating what the candidate is saying as well as making affirmative, encouraging comments.
10. The Civil Rights Act of 1964 forbids questions regarding pregnancy or future plans to become pregnant as well as children and child-care arrangements.

Chapter Cases Answer Key

Establishing Job Descriptions

Headwaiter

- Responsible for directly supervising the daily operation of the dining room and ensuring that daily schedules and established quality standards are maintained
- Coordination of the individual and collective efforts of assigned staff
- Coordinates with host/ess regarding assignment of service stations to all servers and bus persons
- Oversees training of all new service employees

Server

- Responsible for coordinating entire station and communicating with front and back of the house to provide dining experience that meets or exceeds guest expectations
- Processes guest order s to ensure all items are prepared properly in a timely fashion
- Observes diners to ensure that guests are satisfied with food and service, responds to additional requests and determines when the meal has been completed
- Totals bills and refers guests to cashier. May assist bus person with stocking, removing and resetting dishes and flatware between courses and cleaning resetting vacated tables

Cashier

- Ensures an accurate accounting of all transactions, collections and disbursements Receives cash drawer at the begging of shift and counts money at the beginning and end of shift to verify its accuracy
- Itemizes and totals food and beverage checks into register
- Collects cash, check or credit card payment form guest and makes change for cash transactions, check identification, and prepares voucher for credit card purchases
- May reconcile checks, cash receipts, and charge sales with total sales to verify accuracy of transactions at the end of a work shift
- May greet guests, taking food and beverage orders and giving them to the kitchen staff

Host/ess

- Coordinates with Headwaiter in assignment of service shift stations to servers and buspersons

- Informs service personnel of daily menu changes and specials and other matters affecting dining room operations
- Adjusts patron seating flow to balance customers among the various service stations
- Graciously greets guests upon arrival, escorts them to their table, provides menus for them once seated, and informs guests of the name of their assigned server
- Manages special seating requests of guests consistent with table availability
- Relays messages to bus persons and servers as appropriate
- Assists with the duties of servers and bus persons as needed to maintain service effectiveness and efficiency

*The previous job descriptions are paraphrased from the National Restaurant Association Model Position Descriptions for the Restaurant Industry 2000 published by the NRA, Washington, DC

Unwanted Suitor

The manager should step in as Bob's behavior may constitute sexual harassment by creating a hostile work environment. Bob should be spoken to by the manager. The manager should explain appropriate behavior at the workplace and that Bob's persistence is threatening Christine's employment status, as she is considering leaving. Were she to leave due to his continued inappropriate behavior, his own employment would be terminated. If he does not feel capable of ceasing the behavior, then his employment will be terminated. It is the duty of the employer to maintain a work environment free of harassment.

Chapter 2 Test

True/False

T F 1. The key to conducting an effective interview is to remain loose and open to the direction the conversation is heading.

T F 2. Candidates who do not answer interview questions spontaneously, but rather need time think, have probably got something to hide.

T F 3. In order to maintain sanitary standards servers hands or fingers should not touch the inside of cups, glasses, silverware, or dishes where food or liquid or lips will touch.

T F 4. Most servers lose their jobs because they lack the necessary skills

T F 5. Dram shop laws hold anyone serving alcohol liable if an intoxicated person injures or kills a third party.

Multiple Choice

6. Which of the following is NOT a reason for a restaurant manager to hold interviews for server positions?
   a. In order to gather applications to keep on file for the next time that there ARE openings
   b. To gather information about skills, personality and job knowledge
   c. To gather specific information about the particular candidate being interviewed based on data in the application
   d. When an individual is being seriously considered for the position

7. The purpose of asking open ended interview questions rather than closed end is to:
   a. Keep the time required to conduct interviews to a minimum
   b. Gather interesting personal information about interviewees in case you would like to get to know them "outside" of the job
   c. Get much more helpful information in estimating a candidate's suitability for the position.
   d. Establish from the start that employees need to listen carefully and think before they speak

8. Why is it recommended to sit next to the candidate during an interview rather than having a desk or table in between?
   a. It is NOT recommended to sit next to candidates. The desk serves as a way to provide an area of boundary and comfort between the boss and potential hire
   b. In order to help put the candidate at ease
   c. So that the potential employer may be able to gauge any hygiene issues that the candidate may have
   d. In case the employer needs to jump up and answer the door or attend to problems

9. Which of the following questions is a legal and acceptable question to ask of job candidates for the position of server?
   a. What country did that name come from?
   b. Do you own a car?
   c. Have you been convicted of a crime?
   d. How tall are you?

10. Which one of the following is a recommended practice regarding employee uniforms?
   a. Providing one uniform to each employee
   b. Having pocketed shirts in which to carry pens and corkscrews.
   c. Have workers change into their uniforms at work if possible
   d. Require uniforms that emphasize the appealing physical features of the server

11. One of the most important traits a server can possess is:
   a. A positive attitude
   b. A strong back and arms for carrying trays
   c. The ability to talk about themselves to guests
   d. The ability to carry plates on their arm and pour hot and cold beverages from great distances

12. Servers who make mistakes should:
   a. Cover them up
   b. Blame others
   c. Force a smile and find a comforting platitude to say to the affected guest
   d. Admit the mistake and correct it as quickly as possible

13. Which of the following forbids employers from asking non-job-related questions that might discriminate against a group of qualified job applicants?
   a. Family and medical leave act (FMLA)
   b. Fair Labor Standards Act
   c. Privacy Act
   d. Civil Rights Act

14. Which of the following requires employers with 50 or more employees to offer up to twelve weeks of unpaid leave in any twelve month period for birth, adoption, care of spouse or own serious health condition?
   a. Family and medical leave act (FMLA)
   b. Fair Labor Standards Act
   c. Privacy Act
   d. Civil Rights Act

15. Which of the following protects workers between the ages of 40 and 70 from discharge because of age?
   a. Family and medical leave act (FMLA)
   b. Fair Labor Standards Act
   c. Privacy Act
   d. Civil Rights Act

Short Answer/Essay

16. Describe methods by which servers acquire product knowledge. Explain why it is important to do so.
17. How can management keep employees from unionizing? Isn't it harder to maintain control?
18. Can an employer be held responsible for the behavior of guests if an employee accuses them of sexual harassment? What can the employer do to avoid such claims?
19. How can an operation avoid serving alcohol to minors? Intoxicated patrons? Habitual alcoholics?
20. Whose responsibility is it to determine the legal status of employees to work under the Immigration Reform and Control Act? Why?

Chapter 2 Test Answer Key

True/False

1. False. The key is to plan your part of the interview in advance. By preparing questions and structuring the direction before the meeting, you are more assured of getting valuable information than by simply winging it.
2. False. Candidates may need to think out their answers. This is not a clear indication that something is wrong but that the candidate is thoughtful.
3. True
4. False, most servers lose their jobs largely because of personal traits or negative attitudes
5. True

Multiple Choice

6. a
7. c
8. b
9. c
10. c
11. a
12. d
13. c
14. a
15. b

Short Answer/Essay

16. Servers acquire product knowledge through training, studying the menu, knowing how items are prepared and what ingredients they contain; knowing about specials before the shift begins, tasting unusual dishes and asking when they are uncertain.
17. Managers must be careful not to do anything illegal in regards to the rights of unions to represent employees. Management loses a level of control when employees are represented by a union as the union represents the interests of the employees in regards to disciplinary actions, job descriptions etc. The best way not to have employees join a union is to treat them in a manner whereby they do not believe it is in their best interest to be represented by a third party.
18. Yes, sexual harassment includes conduct that unreasonably interferes with work performance by creating an intimidating, hostile or offensive working environment. The employer can begin by not choosing employee uniforms that are suggestive or revealing and may be viewed by guests as an invitation to harass servers. The tone of the establishment in terms of humor, treatment of guests, entertainment and décor can also give a message that employees are valued or not as people. Servers need to know that management is always on their side when guests behave inappropriately and that the manager will step in and take over when a guest has behaved badly.
19. An operation can avoid serving minors by having specific steps in place to verify identification. Documents without photographs cannot be accepted. If in doubt, examine the document carefully. Be watchful for those who are of age purchasing or attempting to purchase alcohol for minors. In the case of intoxicated and alcoholic patrons, refusing

service can be challenging. Servers must be trained to interpret guest behavior and monitor drinking. If an intoxicated guest tries to drive away, management must be informed. If all else fails, the police must be called.

20. The IRCA of 1986 makes it illegal to hire aliens not authorized to work in the United States. Employers must verify citizenship at the time of hiring. Aliens hired before November 7, 1986 can be kept on the job due to a grandfather clause. Every new employer must complete a Form I-9 upon hiring and put it in the employee file. The responsibility is the employers as they are required to only employ those legally able to work. If employers were not held responsible for having underage, undocumented or otherwise illegal persons working for them, they might not exercise due diligence and be tempted to mistreat those whom they were in a position to threaten with exposure.

Activity/Training

- If the class is part of an operating dining room, students should be practicing with a guinea pig group or with each other. Having students who have served in the past to assist in coaching those who are trying it out for the first time is helpful. Guest comment cards allow students to see what they may have done well and areas of development. It also demonstrates that guests really do notice when ladies are not served first, beverages served from the left etc.

- Having students conduct mock interviews can also be helpful. The late, great Michael Hurst told a story about interviewing servers. He said that he always asked prospective servers what the funniest thing was that ever happened to them. If they could not think of anything, he did not hire them. Those who were able to think of something and then told the story, using their hands to while talking, were, in his experience, going to make good servers. He stated that the ability to laugh at oneself and to communicate expressively translated into some of the same skills needed to describe menu items in a way that sells and to relate positively and effectively with guests.

Chapter 3: Exceeding People's Needs

Objectives

- Explain the steps in resolving customer complaints in order to satisfy guests
- Describe ways to manage service to customers with special needs

Outline

I. Introduction

- It is important to create a guest focus that puts guests first
- A 2004 NRA survey found 74% of guests gave service a good or excellent rating at high priced restaurants
- Ask students what they think the satisfaction factor at value priced establishments is. Discuss reasons why value satisfaction is lower. (Entry level employees at entry level pay, fast pace, variety of customers may have varying levels of courtesy, lack of motivation to give higher level service etc.)
- To avoid complaints, customers should not be:
    - Ignored, rushed or kept waiting without explanation
    - Served anything substandard
    - Allowed to sit at a dirty table or be in unsanitary conditions
    - Be treated rudely or with indifference
- Would these factors apply regardless of category of establishment? (Of course)

II. Managing Guest Complaints

A. Complaints as Opportunities

- Guests often complain with good intentions for which managers should be grateful
- Ask students if they have ever felt their legitimate, reasonable complaint was ignored. What was the result of this? (Likely yes, they felt unhappy, upset, did not return etc.)
- Complaints should not be looked upon as an attack or be taken personally

B. Some Don'ts in Dealing with Angry Guests

- Don't respond defensively
- Don't assign blame (servers represent entire establishment)
- Don't make promises you can't keep

- Don't keep phone guests on hold
- Don't refuse a guest's request without an explanation
- Why might server be tempted to do all of the above? (Gratuity- it's not MY fault, someone/thing else that I have no control over etc.)

C. Taking Action

- When guests become irate, stay calm and professional
- First rule is to listen. Guests want to be heard and understood.
- Accept responsibility (for the establishment) if appropriate
- Ask for suggestions and reach a mutually acceptable resolution
- Apologize for the trouble guest has experienced
- If unable to resolve, find someone who can
- Complimentary menu item or meal, if called for is an investment in return business and positive word of mouth
- If guest becomes abusive or out of control, call management, security, or police
- Ask class for examples of successful and failed complaint resolution and look for common denominators

D. Why Guests Complain

- All complaints are legitimate to the guest making them
- Exhibit 3.1 shows examples of why guests complain.
    - #1 is poor timing/flow of service-Discuss what this means and how management might improve this aspect
    - #2 is food quality and temp
    - #3 is sanitation-Discuss guest awareness and importance of ServSafe or other sanitation training for staff
    - #4 is staff friendliness and product knowledge-Discuss managements role in improving this aspect
    - Other reasons are varied
- Guests disposition is also a factor- bad mood, ill, or bad day can all affect perception of service
- Emotions are closely mixed with food which provides comfort
- Servers must be trained in complaint handling as well as understanding how to soothe those whose real trouble is unrelated to the meal
- Patience, tolerance and professionalism are necessary

E. Signs of an Unhappy Guest

- Irritability
- Avoidance of eye contact
- Looking around the room (for someone to help or notice)
- Not eating or drinking items that have been ordered
- Suddenly asking for the check

F. Avoiding Complaints

- Adherence to service standards helps avoid
  - Take orders carefully (to avoid mistakes)
  - Communicate thoroughly with guests (preparation times, etc.)
- "Challenging" guests should be handled with special care
- Complaints due to kitchen action should be brought to kitchen attention immediately and corrected, if possible before return to table
- Would kitchen be more prone to respond positively if their pay was partially dependent on guest satisfaction? Discuss

G. Keeping Managers Informed

- Managers should be informed of all complaints even those that have been successfully resolved
- Managers should always be called in difficult situations
- Management must investigate problems and provide solution that will prevent further problems from occurring
- Managers apology is always good policy and welcome
- Why does having manager apology benefit guests? (know problem has been taken seriously

III. Serving Guests with Special Needs

A. People with Disabilities

- Crutches and canes should be stored near the table but not in the way or under the table
  - Armchairs are helpful for guests who need to raise and lower themselves into the chair
  - When guest is ready to leave, the walking aids should be brought quickly
- Guests in wheelchairs may only need to have the chair removed

- If the guest requests, the server can help move them from wheelchair into dining chair
    - Wheelchair should be stored out of the way
- Disabilities of the arms or hands
    - Some may only be able to eat with spoons
    - Tumblers are preferable to stemmed glassware
    - Hot beverages should be served in a deep cup
    - Some guests might carry their own eating utensils
    - Some of these same principals apply to families with small children-discuss being sensitive to individual needs and mindful of dignity of every guest (hospitality)

B. Communication Issues

- Unconventional communication may include nodding or pointing
    - Pad for writing might be helpful
    - A companion may be the spokesperson for the guest
    - Should the guest then be ignored? (Continue to make eye contact and address questions to the guest while third party assists)

C. People with Vision Impairment

- Greet the sight impaired guest each time you approach the table to announce your presence
- Offer to lead the sight impaired person to the table but take cue from guest
- Pull out the chair and put the hand of the person on the back of the chair to get their bearings
- Guide dogs are allowed in food services
    - They will lead their owner to the table, following server
    - Lead the dog to a place near the table out of the way but where it can see the owner or at the owners feet
    - Do not pet, feed or disturb the dog
- When presenting the check announce the total and go over individual prices
    - If using a credit card, restate amount and help guest find signature line
    - If guest pays with cash, state the bill denominations and coins of all change

D. People with Special Dietary Needs

- Aging population increases dietary needs
- All people are becoming more health conscious

- Requested items include low-salt, low-carb, low or high fiber, low-cholesterol, low calorie or high protein
- Ask students about other requests they are familiar with (Vegetarian, vegan, raw food etc. may all be named also lactose intolerant or food allergy)
- It is not the mission to enforce healthful eating but to make a variety of foods available to satisfy simple needs and tastes
- How have films such as Supersize Me and Fast Food Nation drawn attention to our industry. Discuss
- Servers should know enough about common needs and menu items to assist guests in making selection
- Servers should have a good relationship with cooks/chefs in order to ask questions and communicate requests
- Managers should reward servers who help guests and keep them coming back

Chapter Review Questions Answer Key

1. A guest who was deaf or hard of hearing would be served by; facing the guest and speaking clearly, confirm the order by pointing to the menu or repeating the order back to the guest.
2. A guest complaining about slow service and poor food might be handled by; listening to the guest and taking the complaint seriously. Ask the guest for suggestions in solving the problem and work to find a mutually acceptable solution, finally offering a sincere apology. This complaint presents an opportunity to solve the problem so that other guests do not experience these difficulties and the disserved guests may feel satisfied and return.
3. Serving a blind guest would be done by greeting the guest each time the table was approached, thus announcing your presence. The guest should be offered to be led to their table by offering an arm, but following the cue of the guest as to the acceptance or denial of the gesture. At the table, pull out the chair and put the hand of the person on the back of the chair to locate it. Guide dogs are allowed in food service. Do not pet, feed or disturb the dog. When presenting the check, clearly state the total and go over individual prices. If using a credit card, the server may have to help the customer find the proper line to sign. Change denominations for cash payments should be stated.
4. The three steps in receiving and accepting a complaint are:

   Step One: Listen to the guest and find out the real cause of the complaint

   Step Two: Ask the guest for their suggestion and work on a mutually agreeable solution

   Step Three: Apologize for having caused the guest any trouble
5. When tired parents with crying children have been seated the outside circumstance that may cause the guest to complain is the fact that they are tired and their children (presumably hungry) are crying. The way to avoid a complaint would be to promptly attend to the table, providing some activity to occupy the children. Ask the parents what can be brought

immediately. Generally in this sort of situation, some of the normal rules of serving etiquette are abandoned in favor of quickly getting the kids settled down and fed. When in doubt, always ask the parent what can be done to best serve the family.

6. Crutches, canes or walkers can be stored near the guest's table but not in the way, or directly under the table.
7. Guests in wheelchairs only require that the server remove the chair at the table so that the guest can pull up. If the guest requests it, the server can help the guest move to the chair and store the chair out of the way.
8. Guests with speech impediments may be courteously asked "Did I understand you to say…?" to clarify. Guests who speak little or no English may be assisted by a staff member conversant in their language. Body language, pointing and nodding can help. Have guests point to items on the menu if they show an ability to read more English than they can speak.
9. Complaints improve business because they bring things to light that may also be dissatisfying to others. It is wise not to lose current customers due to the difficulty in attracting new ones. Word of mouth from unhappy guests is destructive. It is better to solve the problem for the current unhappy guest and future guests than pay to advertise for new customers, sending some away unhappy.
10. False. It is nearly always appropriate to remove a charge for an item that the guest has found to be unsatisfactory, particularly when evidence of a problem exists. When a guest has had a long wait, sending out a complementary appetizer, beverage or dessert may be just the thing to compensate the guest for being disserved. An entire meal may be appropriate if the restaurant has made serious errors. Though people sometimes complain in order to receive free items, it is better to risk this as an occasional occurrence than lose guests with legitimate complaints with the assumption that they are scamming.

Chapter Cases Answer Key

Establishing Rules

Student answers will vary but should follow guidelines presented in the appropriate chapters. Chapters Four and Eight has guidelines on check handling. This chapter addresses complaint handling and student rules should follow the three step guidelines, perhaps addressing the don'ts. Emergency situations are also addressed throughout the text with a fire plan outlined in conjunction with the section on French Service in Chapter Seven. This chapter lists serving guests with disabilities, the previous chapter alcoholic problems and subsequent some of the others mentioned. At any rate, the idea is to get the students discussing these issues and the reasons behind the rules.

* Note- it may be appropriate to assign "**The Roast Beef Incident**" case from page 87 here as it involves complaint handling.

Lynne should have begun by following procedure and not taken an order for MR Roast at that point of the evening. The kitchen should also have caught the error and reminded Lynne. She could have suggested another cut of beef that could be prepared MR. Failing that, Lynne should have informed Mrs. Wright of the doneness issue and let her decide if she would accept the beef (apparently she would have) or would like another cut. The host should have accepted Mrs. Wright's decision but may have thought she was simply being polite and wished to intervene on her behalf. The host may have made the error of being overbearing, but this may have been stimulated by the perception that Mrs. Wright was being disserved. The manager went overboard in removing all charges and offering free wine for the table. It probably would have been enough to offer to remove the charge for the roast beef dinner or provide a bottle of wine for the table, if they were drinking wine. Giving the table a new server may have allowed Lynne to be relieved from a situation that had gotten somewhat out of control and given the appearance of solving the problem. It also may have helped Lynne to "get back on the horse that threw her". As the table was now appeased, she probably could have resumed service. Lynne needs to be reminded of the procedures and the importance of guest service. She is a generally a good service, admitted her errors and went to the manager when the she saw the situation leaving her control. Retraining or simply coaching should do the trick. She may just have been having a tough night due to being tired. Firing her would surely be a mistake. A warning should be plenty as she seemed well aware of the consequences of her actions.

Chapter 3 Test

True/False

T F 1. In order to keep guests from waiting too long for a table, guests who are already seated may need to be rushed to finish and allow others the chance to be seated.

T F 2. Complaints are often attacks on the operation and scams designed to elicit an offer of free food. These sorts of guests are the kind most restaurants are better off without.

T F 3. Emotions are closely mixed with our food. Eating provides comfort.

T F 4. Animals, even guide dogs, are never allowed in food service establishments due to sanitation concerns.

T F 5. Food service establishments are subject to special requests from guests wishing to meet special dietary needs due to the aging of the population and a concern with health consciousness.

Multiple Choice

6. In a food service establishment, the focus should be on:
   a. Satisfying the kitchen crew. It is hot in the kitchen and they are working too hard to suffer nonsense from customers or servers.
   b. Satisfying the owner. The business belongs to them and what they want is of primary importance.
   c. Satisfying the bank or lenders that are making it possible to have a food service operation.
   d. Satisfying and pleasing guests from the moment they enter until the time they leave.

7. To avoid customer service problems, servers should always:
   a. Keep customers waiting without acknowledgment and explanation- it will simply sound like an excuse.
   b. Serve guests in the most efficient and effective way possible
   c. Allow guests to sit at a dirty table so they feel they are being seated as quickly as it was possible to seat them.
   d. Rush customers who are lingering beyond the time management wishes them to stay. This makes way for more paying guests or allows servers and cooks to head home.

8. Complaints should be looked upon as:
   a. Attacks on the operation. Loyal managers and servers should defend the operation and its policies to ungrateful guests.
   b. Scams designed to get something free from the operation.
   c. Part of doing business. Some people are never happy, no matter how hard you try.
   d. Opportunities to improve by pinpointing problems that can now be solved.

9. All but one of the following are "Don'ts" in dealing with Angry Guests:
   a. Don't give in to guests who want something free

b. Don't be defensive
c. Don't assign blame
d. Don't keep phone guests on hold

10. It is preferable to have:
    a. A silent dissatisfied guest who simply does not come back
    b. A complaining guest
    c. Guests who are taking a bad day or mood out on the server
    d. Any kind of word-of-mouth

11. Which of the following is NOT a sign of an unhappy guest?
    a. Looking irritated
    b. Not eating or drinking after being served
    c. Being so engrossed in conversation that they ignore the servers attempt to check back.
    d. Asking suddenly for the check

12. The best way to communicate with a guest who seems to be having issues being understood is to:
    a. Speak loudly
    b. Pay careful attention and follow cues as to how to best assist
    c. Ignore them to save embarrassment and have their companions do the talking
    d. Give them a Braille menu

13. The best way to deal with guests in wheelchairs is to:
    a. Lift them carefully out of the chair and into their seat before they need to ask
    b. Remove the regular chair so that the guest can remain in the wheelchair unless they ask for assistance being seated
    c. Find an out of the way place to seat the guest, preferably away from the door and restroom
    d. Provide a chair with arms so they can raise and lower themselves.

14. All but one of the following are recommended for serving guests with vision impairment:
    a. Greet him or her each time you approach the table
    b. Offer to lead them to the table but take your cue from the guest
    c. Welcome the guide dog with a dish of food and water
    d. State individual prices when presenting the check

15. Which of the following is NOT true when addressing the issue of people with dietary needs?
    a. Servers should know enough about common dietary needs to be able to make suggestions and answer questions
    b. A good working relationship with the chef and cooks is essential in dealing with guests who have special dietary needs.
    c. Having a "no substitutions" policy is a good way to head off special requests that slow production and service
    d. As a person's age increases, special dietary needs increase.

Short Answer/Essay

16. What are recommended actions to take when a guest becomes irate?
17. How can complaints be avoided?
18. Should a manager be informed of every complaint, even those that are handled successfully?
19. Why is the food service industry being asked to participate in providing a healthier diet for customers?
20. What can be done to anticipate the needs of guests with special dietary requests?

Chapter 3 Test Answer Key

True/False

1. False. Seated guests should never feel that they are being rushed to leave. To do so, is to cause a service problem. These guests may well leave earlier but may tell others about their unpleasant experience.
2. False. While some persons do attempt to defraud, most are complaining with good intentions. Wise managers listen to complaints and try to solve the problem for the current complaining guests and future guests who may be affected. It is more cost efficient to retain current customers than attract new ones.
3. True.
4. False. Guide dogs are allowed in food services.
5. True.

Multiple Choice

6. d
7. b
8. d
9. a
10. b
11. c
12. b
13. b
14. c
15. c

Short Answer/Essay

16. Servers must stay calm and composed and detached from personal negative feelings, staying professional at all times. Listen accept responsibility if it exists and resolve the problem if possible. Servers should try never to let a dissatisfied guest leave angry. A free item or meal may be offered by the manager if the server is unable to resolve the issue. If the guest becomes abusive or violent it is best to call management, security or the police.
17. Complaints may be avoided by having and adhering to service standards. If guests appear to be unhappy at the start of the meal, sensitive servers can take special care to satisfy them.
18. Yes. This way, corrective action can be taken to avoid recurrence of the problem. A manager's apology is always good policy and welcome.
19. More and more guests are eating out more often and are becoming aware of the indulgent nature of meals out that were formerly reserved for special occasions. The food service industry is responding by making a variety of foods available to satisfy simple tastes and needs.
20. Items on the menu may be labeled to be appropriate for common dietary requests. A statement might be made regarding what sort of modifications are or may not be available.

For example, it may not be possible to prepare a low fat cream sauce dish or low salt dish when bacon is an integral part of the recipe. Thoroughly training servers regarding dietary issues and adding commonly requested items to the menu make items in demand available without being earmarked "special" (Other answers that take on a guest friendly, problem solving approach could work as well)

Activity/Training

Have student's role play the following situation:

A guest has requested that they receive a broiled or grilled version of a fried entrée.

Result #1– The server has promised that it is possible and is basking in anticipation of a large tip. The kitchen is faced with "washing" the breading from a frozen item designed to be deep fried. The server insists and the item is served to the guest who is unhappy with the outcome and refuses to pay.

Result #2– The server informs the guest that s/he will ask the kitchen if it is possible. The kitchen says it cannot be done but provides some alternative suggestions for the server to take to the table. The party agrees to an alternative and is pleased with the outcome.

Result #3– The server informs the guest that s/he will ask the kitchen if it is possible. The kitchen says it cannot be done but provides some alternative suggestions for the server to take to the table. The guest is angry that their request is not being honored and demands that it be done or they will walk out without paying for the appetizers and wine already consumed.

Chapter 4: Service Mise en Place

Objectives

- Oversee the maintenance of well-stocked and organized service stations
- Explain the importance of mise en place for servers

Outline

I. Introduction

- Many preparation steps must be completed before guests arrive or service becomes chaotic and disorganized resulting in weary servers and dissatisfied guests
- Prework that takes place before service begins is called ***mise en place*** which is French for "put into place". It includes:
    - Getting ready
        - This is work done prior to guests arrival
    - Sidework during service
        - Greeting, seating and providing for guests throughout the meal period
    - Ending the meal
        - Helping guests depart, closing the station and leaving the establishment in good shape for the shift to come
- Mise en place differs for different serving situations; the meal period, formality of establishment and type of service
- Ask class if they have worked in or experienced as a guest, situations where mise en place was or was not well done. Ask them to speculate on why it doesn't always happen. (Poor management, understaffed, poor forecasting etc.)

II. Getting Ready

A. First Impressions

- Outside and surrounding grounds must be inviting and give a favorable impression.
- Signs should be well lighted and visible from some distance so guests can be prepared to stop
- If valet parking is available, it should be prompt with neatly groomed and uniformed, courteous employees who assist guests exiting their vehicles

- Ask class if they have experienced valet parking. What is the purpose and where might it be employed? (Places with limited parking, an additional service of a fine dining establishment, a hotel or resort… Valet parking can help to entice guests in areas that are difficult to find parking in or where guests may have to walk, pay meters or parking facilities etc. such as cities, or other areas. Restaurants that lack adequate parking may be able to lease nearby space and remove cars to there for guests so that guests would not otherwise turn away from a full lot)

  1. Lighting

     - Employees and guests need to see well in parking lots, walks leading to front and backdoors, in stairways, near entrances and in vestibules
     - Lighting serves as a deterrent to crime. Criminals avoid well lit establishments in favor of one with dimmer lighting to avoid detection
     - Lighting is a key element of the first impression for entering patrons
     - Quick-service, and employee cafeterias strive for feeling of brightness and movement to contribute to higher guest turnover
     - Fine-dining restaurants need less light to enhance quieter, slower paced relaxation
     - Guests should have sufficient light to read the menus.
     - Candles or burning lights must be properly protected and used with care
     - There must be enough light in the kitchen so there are no shadows on work places. Fluorescent is the most uniform and lowest energy cost
     - Light bulbs should be replaced on a regular basis while they are still giving adequate light, changing all at the run-out time. This saves money over waiting to replace lights singly as they burn out.
     - Ask class to guess what lighting would be like at a variety of establishments. Ask for their own lighting preferences generally and in the following situations; out with pals for a bite, romantic date, meeting with employer, taking toddler out for an ice cream cone, etc.)

  2. General Security

     - Décor is important but should not interfere with safe passage through the establishment

- Telephone should be available with phone book where guests may make calls. Area should be well lit.
- Might it be a good idea to have an area for guests making and receiving cell phone calls? Discuss feelings on cell phone use in quick-service, fine dining, and other situations. (Class will have a range of views but in fine dining, making and receiving calls is generally considered to be rude) Should restaurants block calls or ask guests to turn phones to vibrate mode? (Depends on market, liability if emergency call is not received? Physicians, etc.?)
- Doors and windows should be clean and bright with attractive, well lit vestibule or hall
- Coatroom/rack should be near the entrance
- Management should be alert for criminal activity such as theft by pickpockets etc. Police should be called when suspicious activity is observed

B. The Dining Area

- observant and responsible for their own areas to create an overall favorable impression
- Décor should be fitting to ambiance and clientele. "Less" is sometimes "more."
- Quick-service operations should be clean and have a place for depositing trash and trays
- An inspection before the meal period can ensure that standards are being met and that corrections can be made before guests arrive
- What can places that are open 24 hours per day ("California" style menus) do to ensure that standards are kept up? (Periodic inspections, cleaning during shift, end of shift and during slow times etc.)

1. Environment Control
   - Lighting, air conditioning, heating and ventilation should be comfortable. Heat and humidity problems should be corrected immediately.
   - Air in the dining room should have 50% humidity with a 70 F temperature
   - Lighting, guests and employees give off heat, therefore the dining room must be monitored and adjustments made
2. Station Mise en Place
   - Pantry, storeroom or station supply dishes, flatware and linen

- Servers (and buspersons) set up supplies and buspersons maintain them generally
- Salt, pepper, condiments, and other items also need to be at hand and kept in clean and full condition
- Clean, sparkling dishes are brought to stand with chipped or cracked dishes discarded along with bent or damaged flatware
- Dishes must be air dried for sanitary purposes.
- Spotting occurs due to hard water and can be avoided by using proper detergents. Wiping glasses or dishes to remove spots is permissible
- If class is associated with the operation of a dining room, review contents of storage area, side station etc. Ask class for suggestions for improvements in the current system based on experience as the class progresses.
- Gueridons or Service Carts may also be brought to the table for food preparation, presentation or delivery. These carts should be cleaned and polished
- Equipment; tray stands, high chairs etc. should be clean, neat and in good working order
- Hot and cold food and liquid holding equipment may be operated and must be in keeping with sanitation standards
- Note to class- Temperature for hot food holding according to ServSafe is now 135°F. 140°F is the previously accepted temperature limit for hot food. Cold and raw food remains at 40°F
- Temperatures must be monitored every two hours and corrective actions taken if the temperature is recorded in the temperature danger zone of 40-135°F
- All food items needed by servers; butter, cream, coffee, water, condiments, and garnishes should be on hand and properly stored
- Coffee should be fresh and prepared properly with grounds removed immediately so that bitter drippings do not reduce flavor quality. Brewed coffee should be held no longer than 45 minutes.

3. Table Mise en Place

- Tables should be sturdy and wobble-free, gum-free, clean and level. Any undesirable condition should be corrected before setting the table
- Silencers are felt pads that quiet the noise of the table service and absorb spills. They are set under the tablecloth so the centerfold is in the center of the table

- The silencer is smoothed and the tablecloth is laid over it, hanging down at least eight inches over the table sides. A visual check should be done to ascertain that the cloth is centered.
- A naperone or laycloth if used is placed over the tablecloth.
- Napkin placement may be in the stem of the wine glass, folded in the center of the service plate or on the tablecloth between flatware if no service plate is used. The fold should be facing the guest so that the napkin is easily unfolded.
- Fine dining establishments may hand the napkin to the guest or place it across their laps. Napkin folding may be intricate and add to the décor of the table.
- If no cloth is used, the table should be washed and wiped dry with place maps, if used, placed squarely, two inches from the table edge.
- Table items such as flowers, candles, salt and peppers etc. are then placed on the table, generally in the center
- Flatware is placed with forks on the left and knives (with blades facing inside) and spoons, to the right. A dessert spoon or fork may be placed in the cover center above the napkin.
- The bottom ends of knives and forks should be placed at least one inch away from the bottom of the table.
- Touching any part of the flatware that will touch a guest's mouth is unsanitary. Flatware should be carried on a napkin lined plate, clean tray or a clean service towel.
- Bread and butter plates, if required, are placed next to but slightly above the salad or appetizer fork. Butter knives should be placed horizontally. If a vertical placement is preferred, the tip of the knife should point the same direction as the dinner knife. Both blades should face left.
- The water glass is placed near the tip of the dinner knife on the right. If a wine glass is set, it is placed to the right of the water glass. White wine glasses, (generally the smaller) are placed above the water glass and the red wine glass (generally the taller) is placed slightly above and between the two.
- Many of these conventions may seem fussy and unnecessary to students. Remind them that there is a practical basis for them, such as ease of service and guest comfort level.
- Symmetry is a must as is guest convenience.
- If a table is set against a wall, the items normally placed in the center of the table go in the center, against the wall side of the table.

- Coffee cups may be on the table in coffee shops. Flatware is limited to a knife, fork and spoon.
- In quick and self-service restaurants, the table is bare and guests bring the items they need to the table.
- This would be the time to practice setting up tables if the class is associated with the operation of a dining room. If the class does not have a lab component, students can still identify the type of restaurant based on set up or spot the set-up errors on drawings or photographs of table set ups.

C. Guest Checks

- Computerized Point of Sale systems make issuance of paper checks unnecessary as they are generated by the computer printer with information stored electronically
- Where there is no POS system, numbered paper checks are issued to servers in set quantities or, occasionally, as guests are seated. The form whereby servers sign for checks is called the traffic sheet.

1. Employee Theft

- Employees may be able to steal money through the manipulation of guest checks. Managers should watch for a slower flow of checks, missing checks, fake checks or checks regularly turned in late. Checks can be destroyed or “bunched” and used to receive product, charge and fail to record transaction.
- Other ways of stealing include: giving incorrect change, raising prices or charging for items not served, stealing checks and using them, and taking payment or gratuity money for other servers tables
- Many of these problems can be alleviated through the use of a cashier who rings up the sale and gives change, if any, to the server to return to the table
- An inspection should be made of returned paper checks, with the servers responsible and charged for any missing checks.

2. Guest Fraud

- Guests may steal by “skipping” out on paying their bill or lying about it. They may fail to report a bar bill that was not transferred or alter a bill by erasing or changing entries.
- Careful cash register use, stapling of related checks and retaining control of related checks as long as possible can help avoid these problems
- Customers can claim they were shortchanged so that change making should be done by announcing the amount given, leaving the bill(s) visible and then counting out and announcing the change

D. Getting Servers Ready

1. Scheduling

- Host/ess or maitre d' or other head of service performs mise en place by forecasting server staffing needs by estimating number of covers and dividing by the number one server can take care of during a meal period.
- Servers should be informed of their schedules well in advance so they can plan other priorities
- If someone misses work or the forecast is off, well-organized staff can smooth the unexpected and still have satisfied guests
- Some operations keep a list of extra board service employees or have an on-call system
- Discuss the ethics of having someone on call that must be available to work but is not compensated for that availability. Should the on-call list be voluntary? Should on-call workers receive a stipend? What consequences should there be for an on-call worker who fails to answer a call or is not available?
- Station assignments are often posted by chart which may be restructured or reassigned to allow for station coverage during breaks or the end of shifts
- Does it matter which station a server is assigned to? (Depends, stations may be preferred by guests or considered less desirable and thus less likely to stay full and profitable. Servers sometimes earn stations through seniority or other merit systems.)

2. The Preshift Meeting

- A short meeting of servers or all employees just prior to the meal period can be called lineup and may include an informal inspection for readiness
- The purpose is to go over the menu and service details such as "86's", soups or specials of the day and information about reservations
- All pertinent information should be covered including what goes with dishes, what manner of service, price, preparation time etc.
- This information may be also or only posted on a bulletin board or central information area in the kitchen
- Complaints or compliments are sometimes received but, due to the stress that may follow, it is important to end the meeting on a positive, motivating note

III. Clearing and Resetting Tables

- Clearing and resetting must be done promptly and properly to maintain a good image and quickly accommodate waiting customers
- The linen may be changed in a manner that does not expose the bare tabletop in a variety of step-by-step methods in fine dining situations
- Uncovering the table in order to change out linens is routinely done for large volume operations
- When a bare table is used, it is wiped and washed according to sanitary standards

IV. Ending the Meal

- Mise en place is required to prepare for guest departure. The bill should be totaled and delivered either face down in front of the host or at the center of the table when host is unknown
- A pen is supplied for credit card payment with a rapid return for the processing of cash or credit transactions appreciated
- The server and all other staff the guests encounter upon leaving should offer a sincere thanks and a hope that the guests return soon
- Servers should complete Sidework as the shift ends
- Soiled dishes and linen should be taken to the proper places and clean stock replenished
- Condiments should be filled, cleaned and checked for cleanliness with functional maintenance performed if necessary of dispensers
- Health regulations and the sanitary principles that guide them must be followed. Perishable condiments must be refrigerated; items not in sealed containers that have been put on tables must be discarded. Sealed items are reusable.
- Closing chores differ by establishment but setting up dining and service areas for opening or the next shift generally involves working through checklists that outline duties.
- Stations should be left clean, well stocked and prepared for the next shift.
- Why not do the cleaning at the beginning of the next shift when everyone is not so tired? (Sanitation, mise en place)

V. Quick-service Mise en Place

- Attention to detail is as important in quick-service as fine dining
- Counter and floor employees are the primary servers with orders repeated, verified and paid for prior to the food being served.

- When the order is assembled and placed in tray or bag, the guest typically takes it to a table or service area for condiments and utensils, Guests sometimes bus their own tables or this is done by floor employees.
- In drive-through operations, employee helps guests select items, takes and places the order, expedites and gives a courteous positive impression to the guest
- Drive through requires precise and considerable mise en place due to long lines and quick service times. Servers must work quickly and accurately- thus mise en place is considerable to support this.
- Bags, wrappers/containers, stirrers, straws, cream, sugar, napkins and condiments must be on hand. Cash drawers must always contain ample change
- Discuss experience with drive-throughs that seemed ill prepared or where the order was not right. What were the consequences? (guests do not return, tell others not to go- often problem is not discovered until driver is down the road)

VI. The Cash Bank

- Cash banks carried by servers or cocktail servers allows quick payment of the check and reduces work required by the staff
- Servers must provide their own bank or obtain one upon arriving for their shift for making change. They must verify the accuracy of the bank. If the bank is located in a register drawer, server will have a pass code
- Charges to accounts are done in private clubs and hotels. Hotels require that the server identify the guest with a key or key card.
- Credit card use is common. Servers should know which cards are honored and how to process payment
- Corrections on ring ups are required for over rings using a void system. A coupon or complimentary system may require a check adjustment. Common or individual record sheets may be used for recording purposes.
- At the end of the shift, the register will give a total printout of transactions made by each server. The server's records any information required and turns in the cash owed, keeping the remainder as tips.
- What experience do class members have using cash banks as servers? What system do they prefer? Would class members rather pay the server or a cashier? Why? Does it matter what the setting or situation is? (Variable answers)

Chapter Review Questions Answer Key

1. Answers will vary. Having an inviting first impression will entice guests. Grounds should be clean and in good order, landscaping attractive and walkways uncluttered and in good condition.

2. Valet parking, if available, should be prompt and courteous with neatly groomed and uniformed employees assisting guests exiting from their vehicles. It would be expected that Valets have good, clean driving records and integrity as they will have access to guest vehicles.

3. The busperson may help in restocking supplies for station. The busperson works as a team with the server, manager and host to reset tables quickly. Buspersons often assist in clearing between courses and can be alert to needs of condiments on the tables to be restocked during service.

4. Observe station and retrieve what is necessary from the pantry or storeroom to set tables with linen, flatware, glassware, and any dishes that are preset. Condiments need to be clean, full and ready for guest use, whether on the table or in the supply area. Clean dishes should be available on the side stand or supply station to quickly reset tables along with polished flatware. Dishes must be air dried after being cleaned and sanitized but may be polished if spotting is not an indication of dirtiness. Table side carts such as a gueridon and voitures must be readied if used. Candles or flowers, if used should also be in place.

5. A short meeting of servers or all employees is sometimes held just before the meal period. The purpose is to have an informal inspection of the readiness of all stations and servers as well as to go over the menu and any details of service that need to be brought to the staff's attention. This would include items that are "86" or not available, specials and information about parties coming in. Complaints and compliments might be reviewed but the meeting should always end on a positive note to motivate a successful shift of service.

6. Point of Sale systems alert managers to problems by printing checks for each order when a code or password is entered by server, thus recording all sales and voids. Voids often require a manager code. Computers also calculate the check and indicate how much change is due for the amount given.

7. Come to the table with a clean folded tablecloth held under the arm. All items from the table are to be removed to a nearby cart or tray. Place the folded cloth on top of the soiled one. Next, pull the soiled cloth towards you so the hem is even with the far edge of the table. Unfold the fresh cloth by holding the center crease and the top hem then gently flip over the bottom section. Hold the top edges of both cloth and pull in the opposite direction so the soiled cloth can be pulled out while the fresh one remains in place. Make sure the new cloth is centered before making adjustments. Uncovering the table is faster and suited to more large volume operations or casual ones. Move the tableware items back to the center of the table. If bare table is used it must be washed clean and sanitized prior to resetting. Once the table is reset completely, it is ready for the next guests.

8. The bill should be totaled and delivered face down in a tray or holder in front of the host or in the center of the table. A pen should be provided for guests who are paying by credit card. As the guest leaves, they should be thanked and invited to return.

9. Wet dishes or glasses may not be wiped dry following being washed and sanitized as wiping may recontaminate and is often prohibited by health regulations for this reason. All dishes must be air dried. Spot removal due to hard water should be done with a clean, dry, lint free cloth.

10. Answers will vary however the Point of Sale system is the most reliable for recording sales and also has the capability of splitting checks, recording gratuities for IRS purposes, labor hours and other management reports. Other methods have advantages but the POS is the most reliable.

Chapter Cases Answer Key

Good Mise en Place is Missing

Answers will vary but should be along these lines:

- Check all tables and see what is needed to set them up, if they are not already set. If not, obtain silencers, linens, dishes, flatware and glassware as customary to set up the tables.
- Set all tables to the standard of the operation. Glasses and flatware should be checked for spotting and wiped with a clean, soft, lint free cloth if needed.
- Centerpieces and condiments should be placed conveniently for guests.
- Stock side stand or station storage with additional clean dishes, glassware, flatware and linens. Polish and prepare in whatever manner is appropriate.
- Set tables with customary condiments. Most establishments will set out clean and full salt, pepper and sugar packets including sugar substitutes. Additional condiments may include mustard, catsup, steak sauce, vinegar, oil, and etc. all in clean, full containers. Back up for condiments is desirable to have. Filled pepper grinders may be desired.
- Flatware special use items and glassware special use items need to be available if needed such as wine glasses, soup spoons, etc.
- The supply station refrigerator may be stocked with milk, cream or substitutes, lemon, and butter pats. Garnishes and ice should also be hand. Prior to service, such beverages as iced tea, coffee and decaffeinated coffee should be prepared, each held according to recommended times. Hot food should be held at or above 135°F and cold foods at or below 40°F for sanitary purposes.
- Service carts and/or trays should be clean, sanitized and available for use and stocked, if appropriate, with whatever items are customary for service.
- Furniture in the station should be clean and polished. Furniture or related items such as high chairs, booster seats etc. should also be clean

Chapter 4 Test

True/False

T F 1. Mise en place work differs for different serving situations.

T F 2. Outdoor lighting serves as a deterrent to crime.

T F 3. Studies have shown that it is more cost effective to use light bulbs until they burn out and then replace them quickly rather than replacing all bulbs at the estimated "burn out" time.

T F 4. To check to see if a dining room is too humid, fill a glass with ice water and set it down on a dish. If beads of moisture quickly build up on the surface, the room is too humid.

T F 5. As tables are generally clean and bare in Quick-service restaurants, no mise en place is required.

Multiple Choice

6. *Mise en place* is French for:
   a. Table set-up
   b. Put into place
   c. Service cart
   d. The art of changing the tablecloth without exposing the table

7. Bright lighting in places such as Quick Service restaurants serve the purpose of:
   a. Helping employees spot areas that need to be cleaned
   b. Calling attention to the establishment from the parking lot
   c. Achieving a feeling of movement where guests do not linger, assisting in a higher guest turnover.
   d. Reducing crime in the dining room and restroom

8. Dining room arrangements, whether formal or Quick-service should be:
   a. Neat, not too close together and clean. Chairs should be squarely set slightly under the table.
   b. Staggered so as to suggest hominess rather than an institutional orderliness.
   c. Set well apart to allow for privacy and ease of traffic
   d. Set very close together to give a feeling of fun and make more money for the operation

9. Desirable temperature and humidity in a dining room is:
   a. 65°F temperature and 0 % humidity
   b. 75°F temperature and 75% humidity
   c. 68°F temperature and 25% humidity
   d. 70°F temperature and 50% humidity

10. Dishes and glassware that show cracks or chipping and flatware that is bent should be:
    a. Promptly repaired
    b. Set aside for use in emergencies
    c. Set aside to use for "to-go" orders

   d. Promptly discarded

11. "Spotting" on dishes is a result of:
    a. Air drying
    b. Hard water which also causes mineral build up over time.
    c. Dishes dried with a too damp towel
    d. Inadequate washing
12. The following statement about coffee preparation is TRUE:
    a. Coffee should be prepared immediately so that the flavor can build up
    b. Coffee in food service operations is best brewed one cup at a time
    c. Coffee grounds should be kept in the basket until the next batch is ready to brew as a time saving measure.
    d. Coffee should be held no longer than 45 minutes

13. The purpose of a silencer is:
    a. To encourage fine dining patrons to speak more softly so the ambiance of the establishment is enhanced.
    b. To "baffle" the establishment so that the sound of voices is muted
    c. To quiet the noise of table service and absorb spills
    d. To immediately turn off the sound system so that a speaker may address the room or a toast may be heard by all.

14. Guest checks are numbered in order to:
    a. Keep track of how busy the establishment was on any given shift
    b. Keep track of how many tables have been served since the opening day
    c. To ensure that each check is properly collected and accounted for.
    d. To track server productivity

15. Mise en place:
    a. Refers to getting ready for the meal. It has nothing to do with departure of guests
    b. Refers to fine dining only and does not apply to casual operations.
    c. Is entirely different than side work.
    d. Is also required to prepare for guest departure. It includes preparing the bill and thanking the guest as well as inviting them to return.

Short Answer/Essay

16. What is a cash bank how is it used in reference to servers and cocktail servers?

17. When a restaurant is very busy, dirty tables may have to wait until the staff can catch up and get to them. Seating a guest, who has been waiting at a dirty table, lets them know that the establishment is trying as hard as they can to get to them and is accepted practice. Please comment.

18. What kind of fraud have guests been known to commit and how might they be foiled?

19. What kind of mise en place planning is done by the host/ess, maitre d' or service manager?

20. Besides communicating specials and guest forecasts, what is another important purpose of the Preshift meeting?

Chapter 4 Test Answer Key

True/False

1. True
2. True
3. False. Changing all lights at the burn out time saves money over replacing lights singly as they burn out.
4. True
5. False. Attention to detail is as important in Quick-service as in fine dining. Drive through windows, long lines at the counters and the expectation of speed all require considerable mise en place or side work.

Multiple Choice

6. b
7. c
8. a
9. d
10. d
11. b
12. d
13. c
14. c
15. d

Short Answer/Essay

16. A cash bank is when servers of any kind carry their own money. It allows quick payment of a check. The server will make change from their own cash fund obtained at the start of the shift. The bank may be carried or kept in a secure cash drawer.

17. Clearing and resetting dirty tables properly is a high priority and should be tackled by the server, busperson and manager as a team in order to seat and serve waiting guests. No one should ever be sat at a dirty table. It gives a poor impression, is unsanitary for the guest and is inhospitable.

18. Guests may purposely fail to report when their bar bill is not carried over or when an item or items are omitted. They may erase or alter entries on their checks. They may claim they have been given insufficient change. Careful register use, stapling of checks together and keeping control of the check as long as possible can help avoid check fraud. To avoid quick change artists, servers should leave the change on a plate or tray, name the amount and pay the correct change, naming the amount given as change is made.

19. Forecasting server staffing needs based on the estimated total number of covers divided by the number of covers one server can take care of during the meal period. The forecast may be made for a week and daily work assignments should be made so servers know when are to be at work. Servers should be informed of their schedule well in advance so they can plan other priorities. Station assignments are done by chart so servers can begin mise en

place. Stations may be restructured or assigned to allow coverage during breaks or as individual servers end their shifts.

20. Besides communication of menu and guest information the meeting is used to motivate servers and thus should focus on the positive, building the server and the team. This helps to result in great service.

Activity/Training

Have the class, as groups or individuals develop a Mise en place checklist for a specific dining room (provide a menu if it is not one run in conjunction with the class). Include: 1. Getting ready, 2. Sidework during service and 3. Ending the meal. The lists will vary according the restaurant assigned. It may be interesting to break the class into groups and assign various styles of restaurants and compare and contrast the lists. Have class members exchange lists and check for completeness. If the dining room is operated by the class, use the best checklist or a compilation of the lists to create a mise en place checklist for the term.

Chapter 5: Service in Various Industry Segments

Objectives

- Describe proper meal service and clearing for banquets, specific meals, buffets, and other types of service

Outline

I. Introduction

II. Banquet Service

A. Regular Banquet Service

- A banquet can encompass any meal period and be a special celebration, a gathering to honor someone, a professional meeting and may be preceded by a reception.
- Banquets require much planning with entertainment, dancing, sound system, lighting, flowers etc. all arranged in addition to the meal at the request of the customer
- Menu and price must be agreed upon beforehand including service style as specified or planned by the facility. Banquets generally require a special service staff.
- American and Russian service are the most common styles used. In American Service, food is placed on plates and brought to guests, in Russian; the food is dished to guests from platters tableside.
- Room set up may be done by a separate crew or the servers. A sample setup is provided so the model can be followed. Set up is generally done as a team effort rather than the station set up done in traditional dining room service by the individual server assigned.
- Why not have the set up done by station? (While the finishing touches are generally done this way, it is more efficient to do the entire, room at once.)
- Why are uniforms, requirements and set up different each time? (Each function is unique and arranged by a paying host/ess to their specification, making each event one for which servers must be informed anew)
- In standard functions, all flatware, china and glassware are preset. In more elaborate functions, much is brought to the table but this requires more service staff and a higher cost.

- Presetting tableware may crowd the table and intimidate the guest. Presetting may go so far as to include the coffee cup to the right of the guest with the handle at 3:00
- Generally each cover needs 25" of linear space
- Ice water is preset and often the first course, if cold and possibly even bread and butter.
- A lineup or Preshift meeting may be used to give specific instructions on setting up. The level of detail will relate to the experience of the staff.
- Tables are arranged as needed with a head table often facing the room. That table may include a podium for the speaker and is often on a raised platform.
- The most common table arrangement is a 6-10 cover round. 60" diameter accommodates 6 guests and 72" 8-10 guests.
- Squares and rectangles of 6, 8, and 10 feet are also common.
- Round tables are placed 4-5 feet apart and rectangular and other shapes 5-6 feet, though rectangles can still accommodate more people in the same space
- Square, oval, half round and serpentine are used less frequently.
- Why are rounds the most popular? (Rounds allow for easy socializing also, the number of seats can be adjusted and there are no sharp corners for servers to bump into.)
- Tables are numbered so that servers and guests can find them
- Station assignments are by individuals or team. One server can handle 24-32 covers alone depending on the complexity of the function.
- Are there any other systems used besides individuals and wave? (Often a two person "tote and serve" team handles four to five rounds. One person carries trays in and out and the other attends to the tables. The tray carrier "tote", assists with clearing or serving beverages as well. The tote position may be held by the newer team member or physically stronger of the two with the more experienced or gregarious server dealing more directly with guests)
- With the wave system, the entire room is one large station, beginning with the head table whereby servers perform various synchronized functions. This is preferred by many experienced teams.
- Dining space required is 15-18 square feet per guest
- Table cloths are sized and shaped according to commonly used tables; 60" round = 85"x 85" cloth, 8' rectangle = 54"x 120" cloth
- Why all the measurements? (It is essential to be able to predict the feasibility of handling parties of various sizes and configurations. Computer software has been designed for this purpose. If the class is accompanied by a lab it may be

instructive to have the students attempt various configurations of the room and draw up charts of scenarios that work. Mock rooms can also be drawn up with furniture models used to portray various set ups.)

- Various shaped table configurations for special needs include "T," "L," "U," "E," and square.
- What sort of special needs would these sorts of configurations be suitable for? (Meetings where there is a need for everyone to be able to see each other, to maximize seating within a single table, to socialize, etc.)
- Service is simplified since everyone eats the same meal at the same time. Meal service is often brisk to allow for program to follow.
- What if a guest cannot eat something due to an allergy or religious restriction? (Banquet operations generally allow for some special orders. Some registration forms for functions may have guests specify if they are vegetarians for example. The house often keeps simple items on hand, fresh or frozen to be quickly plated, should a guest need an alternative)
- Servers generally never perform service during programming such as a speech, award presentation or guest activity. If an unscheduled toast or announcement is made, servers hold off and wait until it is finished.
- The server must know the menu so they are able to answer menu questions and plan service for each unique function. Complete mise en place is essential for smooth banquet service.
- Pre-plated food may be held in warmers. This simplifies service but may cause food to lose appearance quality
- Dessert service may be simple or spectacular. Cake may be ceremoniously cut, or simple dishes of ice cream brought in.
- Wine or special beverage service may require special pourers to perform this function. Many of the ceremonious steps to wine service are eliminated as the wine has been pre-selected.
- Those hosting the banquet provide an estimate as to the number of covers with an acceptable percentage more or less accepted by the operation. In order to know the actual number tickets, a head count or plate count may be used. Full tables can be multiplied by number of seats with partial tables added.
- What would be the challenges in costing out a banquet? (Estimating no shows or overage- a budget conscious host does not want to pay for more than will actually eat but wants to accommodate everyone. The house wants to charge for food prepared and not have waste while being able to sell as much food as is desired. This is what a banquet contract with a guarantee and agreed upon amount to prepare over are for and why people put RSVPs on their wedding invitations.)

B. Banquet Table Service

- Two predominant service styles in Banquet service are American and Russian with American being the most popular.
- In American service, food is served from the left with the left hand and removed from the right with the right hand, beverages served and removed from the right
- Why are beverages served and cleared from the right? (Beverage glasses are placed to the right of the diner. Serving from the right avoids the need to reach liquid in front of the guest.)
- It is recommended that no more than three plates be carried at a time unless they are small and "dry". Three plates are aesthetically acceptable. One is preferred but slows service.
- When pouring coffee and other beverages the cup or glass should not be removed from the table. To pour over is safer and technically correct
- Does this seem harder? What are the risks? (Students may fear spilling liquid on guests but picking up cups or glasses actually risks spills more. Technique must be practiced but is very satisfying)
- Banquet rooms have more breakage than any other type of foodservice operation. Separating soiled dishes and stacking by size will minimize breakage, speed up breakdown procedures and be a great help to dishwashers.
- Servers normally carry trays on the left hand, flat resting the weight on their shoulders. Experienced servers may balance the trays on their finger tips. Dropped trays result in embarrassment, breakage and possible injury.
- Ask class if they have ever dropped or witnessed a tray full of food or dishes being dropped. (Various- this author experienced a guest standing up suddenly under a full tray of prime rib on metal plates with au jus and was burned from neck to waist- servers need to watch for guests sudden moves or objects on the floor that may cause them to trip)
- It is imperative that servers understand how crucial timing is in banquets. A special request or problem can throw everything off. Teamwork is essential to successful banquet service.
- How much can a guest reasonably expect in a banquet setting? Why do guests make special requests? (Many guests have no idea how their request may be throwing off service. Many are much more health conscious and are not willing to eat something they believe is bad for them. Registration forms or hosts may assure guests that their request can be easily honored. Servers should be aware if special meals have been requested and be ready to accommodate these requests. Guests should be mindful and let the server know early on if they need a kosher, no salt or vegetarian dish)
- Buffet service should be used only with fairly small banquets. If there are too many guests there is confusion and it is difficult to move the line quickly. If it is

used for a banquet the courses may be limited and dessert and beverages may be served at the table by the server.

- How might buffet service be used for a banquet? (Sweet table- discussed later in the chapter with desserts, having salads preset with guests up at a variety of buffet tables for the entrée and separate sweet tables or dessert table service could all be employed. The more guests up and around, the more chaos can be created)

C. Receptions

- Receptions are gatherings to celebrate special occasions or to honor a person or persons they may be small or large
- Often there is a service bar with alcoholic and nonalcoholic beverages
- If guests are expected to carry beverages and appetizers, servers will be needed to take empty plates, soiled napkins and glasses
- Tray stands with trays should be conveniently located out of sight to collect dishes
- Another kind of service is when servers present trays of beverages and appetizers this is called flying service and the trays are then flying platters.
- The Russians influenced the original French reception custom and introduced the concept of standing while eating
- Where would the highest cost be- self service receptions with seating or standing and receiving flying service? (Self-service tends to promote greater consumption, particularly when guests can make themselves more comfortable and sit with a plate. It is hard to have a beverage in one hand and anything more than a single appetizer in the other, thus cutting down on food cost. Even the labor cost of the servers offset the higher cost of guests helping themselves and getting comfortable)

D. Showroom Service

- Showrooms often offer a meal and entertainment at a nominal price to attract patrons to the facility where gambling occurs. This may also occur at a race track or other entertainment venue.
- Beverages only may be offered for a late night show or cabaret
- Showrooms usually include a preset menu, nonalcoholic beverages and entertainment in one cover charge
- Often as many as 500 are seated at once. Serving them requires good organization and planning
- After the orders are taken they are immediately communicated to the kitchen where service moves along due to the limited menu. Items are often proportioned and preset

- Setup procedures for showrooms are the same as for banquets with the room being quickly preset for the next seating
- Have students experienced this type of service before? Can they envision other venues where it might work? (This may be similar to cruise ship service where there are set seating times or other all inclusive resorts, particularly with entertainment)

E. Kosher Meals

- Trained Jewish person is need to oversee complex preparation of food according to Jewish dietary law
- Kosher meals may be found at catered functions such as bar and bat mitzvahs and weddings
- Dietary law strictly forbids that all food prepared with meat products come in contact with dairy products. Kitchens are either separate or have separate sinks, equipment and utensils.
- Kashruth are the laws of keeping Kosher which means fit or proper.
- Laws allow only consumption of meat from animals with split hoofs and that chew their cud. This means pork is forbidden
- Only specific birds, including chicken and turkey are allowed
- Only fish with both fins and scales are permitted. This prohibits all shellfish
- Why do so many religions have laws regarding food and service? (Food is a basic part of survival. It is theorized that some laws were practical in nature, promoting sanitary food, before much was known about how food became unhealthy. Other laws are customs believed to have come from tradition or holy dictate. Regardless of the source, for those who are believers, observance of these laws cannot be taken lightly)

III. Service for Specific Meals

A. Breakfast

- Breakfast service is usually faster than other meals because guests are often in a hurry. (Unless it is a special occasion)
- American breakfast consists of juice, cereal, entrée item, bread and hot beverage
- Continental breakfasts include fruit or juice, coffee or tea, bread product and butter
- Buffet service is often used for breakfast. Sometimes a cook may be there to prepare eggs and omelets to order
- Special occasion breakfasts include:

1. Wedding breakfast which may begin with beverage alcohol and may be quite substantial if served late in the day
2. Hunt breakfast which is an elaborate, heavy buffet with meats and other dishes, often game
3. Brunch is a fairly heavy breakfast served late in the morning which may also serve lunch entrée items, thus the term brunch. Champagne may be served.

- The amount of service for breakfast varies. At buffets, the server may only serve the beverage.

B. Lunch

- Foods served at lunch can vary considerably
- The type of establishment and foods being offered will determine the table setting
- American service is most common for lunch unless the dining is quite formal- the tendency today is to simplify
- Group lunches are often a complete but lighter meal and may include a speaker. The service must be fast as the time may be only one or two hours.
- Service procedure for group lunches is similar to that used at banquets
- What kind of service do people expect when they are picking food up at a restaurant to take away/back to the office? (Similar to drive-through service later in the chapter- courteous, fast, and accurate and complete with needed condiments etc.)
- Is the American lunch hour expanding or shrinking? (Many people are squeezing errands etc. into their lunch hour but others still desire that break in the middle of the day. They want a pleasant and efficient experience)

C. Dinner

- People generally have more time for dinner than any other meal and often want to relax after a day's work
- People often want to dine out to celebrate an occasion and may want a bit more elaborate meal than they would normally consume
- Occasion groups served in the main dining room will receive the same sort of service, if in a special room, it may take on aspects of banquet service depending on the meal and the service desired
- What sort of special occasion? (Birthday, anniversary, engagement, rehearsal dinner, business or personal success, graduation etc.)

D. Tea

- Service requirements will vary with the type of tea (high or low) and the number of people to be served

- For small groups, the service can be similar to serving a family at a meal but for large groups it can take on aspects of a banquet
- Tea is served with milk or cream, sugar or honey and lemon slices on a tray and set on a small table. Cups, saucers, spoons and small tea napkins should be provided
- Server will pour tea and bring it on a tray to each guest
- Guests must be relieved of their tea cups and other items as soon as they finish and the serving table must be kept filled with the necessary items
- Low tea may simply be tea with condiments or may also include sweets, nuts or fancy sandwiches. They may be served seated or standing. Large group teas are almost always standing
- High tea has the elements of the most involved low tea but some heavier foods may be offered, making it a light meal. High teas are always seated service and are not usually practical for large groups.
- Why study tea service? Is this likely to come up? (Tea is becoming increasingly popular and this is the sort of service that is common internationally if not, in the United States. Some hotels, clubs and fine dining establishments do offer tea service. High tea may also include light cocktails)

IV. Buffet Service

A. Regular Buffet

- A buffet is a service concept that presents food on a table. Guests then progress, serving themselves or indicating what they desire
- Buffets vary considerably as to foods offered, when they are used and in the table service required to serve guests
- Guests commonly serve themselves vegetables and side items and progress to the end where a server carves and serves meat or hot items and places a portion on the guest's plate. If the meal is breakfast, the server may make waffles, hot cakes or egg dishes.
- Hot food is served last, immediately before the guest sits down to eat.
- A shopping service is a system whereby the guest goes to various tables arranged with only one kind of food on each table to prevent long lines.
- A sweet table is a lavish buffet of different desserts and may follow a sit down banquet dinner
- In some buffet services, a tray and tray rail on which to slide dishes is provided much as in cafeterias
- Tables may or may not be set with flatware or they may be available for the guest rolled in napkins as part of the buffet line

- The serving side of the table is often protected with a sneeze guard to protect food from airborne germs
- Hot dishes are often kept heated by electric lamps. Plates for hot foods only should be kept warm.
- Cold dishes are kept on a refrigerated tabletop. If cold food only is served, the dishes for service should be chilled.
- All purpose plates should be room temperature
- Advantages of buffets are that they can handle large groups of people quickly if the menu is limited. Service time is also short. They appeal to guests in a hurry. Buffets require a smaller service staff.
- Disadvantages are the higher food waste because guests sometimes take more food than they can eat. Servings may be controlled by proportioning or other portion control methods.
- What kind of portion control methods might "other" encompass? (Having someone on hand dishing out food in portion sizes can inhibit over serving. Other methods include providing smaller salad or lunch sandwich size plates rather than full sized dinner plates. Some buffets offer appealing, filling food at the beginning and middle of the line and save the more expensive fare for a point where the plate is nearly full. Quality bread products served generously can stall second helping time until the stomach has registered full or nearly so.)
- Attractive presentation is important and can be achieved by arranging food in different heights possibly through a raised platform, flowers, mirrors, fabric also add. Colors should be bright, natural and varied. Garnishes should be beautiful but not garish.
- Ethnic Buffets
    - Smorgasbord originated in the Scandinavian countries. A traditional one offers pickled herring, rye bread, mysost or ejetost cheese and Swedish meatballs, among a full spread.
    - Russian Buffet is elaborate and includes rye bread with sweet butter, roasts including game and caviar in a glass or ice carved bowl. Vodka will also be served.
    - French Buffets are not common. They are complicated and feature an appetizer buffet from which selections are made that are brought to the table by servers. This method is repeated with the main buffet. Wine is served. The desserts may be featured on a table or voiture (cart) which is brought to the table.
    - Ask students if anyone has experienced any of these. Students located or from the Midwest may have experienced a smorgasbord. It is more likely they have experienced a Chinese buffet. Ask them some of the characteristics of this kind of ethnic buffet and to explain the popularity.

(Allows people to try different things, value perception, healthy eating perception possible due to vegetables, mainstreaming of ethnic cuisines)

- Action Station consists of cooking and finishing some food items on medium sized pans over portable rechauds as guest go through the buffet line.
    - This is Russian style as the food goes from preparation in the pan to the guest's plate
    - Guests like the fresh preparation and feeling of customization
    - Action stations are typically used for stir-fry, pasta, scampi, tenderloin or other items
    - Particularly with advanced preparations, the server must have a thorough knowledge of the menu and techniques
    - This tableside preparation increases the service level and overall cost of the meal
    - Ask students what other foods may be suitable for Action station preparation (various answers)

B. Cafeteria Service

- Cafeteria service is a type of buffet where the guest selects items desired and carries them away to be consumed elsewhere.
- Foods are arranged in the same sequence as a menu in a typical restaurant, that is, in the order they are usually eaten
- Patrons make a selection and servers behind a counter serve the food which may be proportioned
- Patrons pick up their own eating utensils and beverages using the self-service facilities
- A shopping center arrangement where different foods are obtained from different sections is used as it was in buffets, explained previously
- In some cafeterias, different ethnic foods or regional specialties are obtained from different counter locations
- Payment is made at the end of the line at a cashier stand
- Cafeterias have generally been associated with senior citizen markets and institutional food service. What modern or youthful applications might be made that could revive this concept? (Health food, combination cafeteria/buffet where guests help themselves and then plate is weighed, thus allowing higher quality offerings, Pizza buffet, Super Salad Bar etc.- various answers possible)

V. Other Service

A. Drive-through Service

- Customers drive to an ordering station where the menu is displayed and use a two-way communication system to order
- They continue to a service window to pay and receive their food
- They can then park and eat, drive and eat or take food elsewhere to eat
- This is appealing to families with small children, people in a hurry who get through faster then entering and using the counter, and in the late hours
- Restaurants have extended their hours with drive through and some report as much as 79% of total sales from this popular source
- Some operations have expanded to three stops: order, pay, and pick up which they say speeds service and aids sanitation
- Drive through employees should be friendly and make sure orders are complete as errors will often not be discovered until too late.
- Currently, burgers, tacos and chicken are the standard drive through offerings along with coffee beverages. Can students think of other foods that may be suitable or that there may be a market for drive-through? (Various- healthier choices may be mentioned)

B. Room Service

- Hotels and some motels offer room service in which guests select foods and order them by telephone. These selections are then delivered to the guest's room.
- Breakfast can be preselected the night before using the hanging menu on the doorknob
- Delivery carts for room service should be preset and ready to go. The carts may have means to keep hot and cold foods cold or, more often, covers are placed over food.
- Orders must be carefully checked by servers as, like drive through service, errors take time to correct and are not discovered until too late to quickly correct them
- Guests place most room service orders by phone with the average time span between ordering and delivery needing to be kept to 15-30 minutes
- Some properties use a service elevator for the delivery of room service orders
- Servers should knock and announce "room service" when delivering food, and enter the room, cart first
- The cart is either left to be used as a table or the order is unloaded on the tray into the room.

- The server then has the guest pay or, more often, sign for the charge, leaving quickly to respect the privacy of the guest.
- Why do people order room service- it is quite expensive? (Tired, need privacy, "honeymoon", multi-tasking, expense account etc.)
- What options can a property that does not have room service offer to guests? (Local restaurants that deliver can have menus in rooms and delivery right to the door of the guest. Vending machines on site, sharing a parking lot with a restaurant that may deliver to the lobby, etc. could all be alternatives)

C. Counter Service

- Counter service is fast work under pressure with short turnovers of guests in a hurry.
- Counter stations range from 8-20 covers
- Work areas must be carefully planned so service can go smoothly. Everything should be within reach
- Place mats are often used with condiments close by
- Service can be speeded if order placement and pickup are close
- Cleanup of departing guest and resetting for new guest should be fast- new guests often seat themselves before dirty plates have been removed
- Servers must remain cheerful and attentive to guest's needs despite pressure
- During lulls, Sidework should be done; restocking napkin holders, cleaning and filling condiment containers including salt & pepper and sugar bowls. Lack of adequate mise en place makes busy times chaotic
- Standup counter service is sometimes used to give fast service but often only fast, short order foods are served in this fashion
- Could counter service lend itself to healthier foods? (of course)

D. Fountain Service

- Ice cream parlors experienced a decline in popularity but are now begging to regain ground
- Product usually comes in 2 ½ gallon paper containers and is prepared under good sanitary conditions
- The most common fountain items are; sundaes, sodas, floats, milkshakes, splits and carbonated drinks using standard recipes
- Discuss newer concepts such as Cold Stone Creamery and Kilwin's where mix-ins are custom done on marble slabs and or cones are made in house. Both of these franchises sell retail product, ice cream, ice cream "cakes" and candy

respectively. If the instructor is not familiar with these concepts, call local ones to mind. (Various responses)

Chapter Review Questions Answer Key

1. The different ways in which service can be accomplished is referred to as service styles. (This from page 110) Service styles include; American, French, Russian, English and Chinese. Service styles may also vary according to the occasion, be it a banquet, reception (where flying service might be employed- or cafeteria), Breakfast, Lunch, Dinner or Tea and Buffet service. Still other types of service are used for Drive-through, Room Service, Counter and Fountain Service. This question may be a bit confusing.
2. A banquet is a meal where a group of people are gathered for a common purpose. A banquet can encompass any meal period and may be a special celebration, a gathering to honor someone, a professional meeting and may be preceded by a reception with beverages and hors d'oeuvres. American and Russian are the most common service styles used.
3. Three types of ethnic buffets include the smorgasbord, a Russian buffet and a French buffet. The smorgasbord originated in the Scandinavian countries. If offers a full meal with many kinds of food. To be a typical smorgasbord, pickled herring, rye bread and mysost or ejetost cheese should be offered as well as Swedish meatballs. A Russian buffet is an elaborate offering of hot and cold foods including game. The buffet should offer rye bread with sweet butter and caviar in a glass or ice carving bowl. Vodka will be served. A French buffet begins with hor d'oeuvres and appetizers where patrons select what they want and servers take it to their seats. The main buffet is served in the same fashion from substantial offerings. Wine is served and a dessert table may be featured or arranged on voitures (carts) and brought to seated patrons for selections.
4. A low tea is simply tea with condiments or may include sweets, nuts or fancy sandwiches. A high tea is like the most elaborate low tea but some heavier foods may be offered. High teas are always seated service and not for large groups. Low tea may be served standing or seated.
5. Receptions are gatherings to celebrate special occasions or to honor a person or persons. They may be small or large and preceding a banquet. Often there will be a service bar or table for alcoholic and nonalcoholic beverages.
6. A flying platter is service whereby trays or beverages or hors d'ouevres are presented by servers who are said to be performing flying service. This is typically used at a reception, where servers walk about among standing guests, offering them food and drink.
7. A buffet is a service concept that presents foods on a table. Guests progress along the length of the buffet serving themselves or indicating what they desire to servers attending the tables.
8. Variations occurring in cafeteria service include: a single counter whereby food is arranged in the same sequence as a menu- the order generally eaten with appetizers, salads, pastas, vegetables, main items and desserts. Servers behind the counter serve the foods selected by patrons. A shopping center arrangement is one where different groups of foods are obtained from separate sections to avoid long lines. Payment for food sis made at the end with servers or buspersons clearing and cleaning tables. Service may be provided on special requests.

9. Room service is offered by hotels and some motels in which guests select foods and order them by phone. The selections are then delivered to the guest's room. Special challenges of room service include; keeping food at desired temperature, making sure nothing such as condiments or utensils are missing when food is delivered, time between ordering and receiving food, and the fact that the server is entering the guest's private room. The cart either becomes a table and is left in the room or is unloaded and leaves with the server. The check is left, and almost always simply added to the room charges. If the gratuity is not included, the server is then tipped before leaving.
10. In drive-though service, the food order is placed at an ordering station by the guest from their car using a two-way communication system. The car continues forward to the service window where they pay for and receive their order. Some establishments have added an additional window for payment only followed by a pick-up window which they say speeds service.

Chapter Cases Answer Key

Managing a Menu

Answers will vary- you might have students look on the Lettuce Entertain You website: http://www.leye.com at RJ Grunts menu or the Heartland Café menu: http://www.heartlandcafe.com. These menus have been popular with college aged patrons.

The Roast Beef Incident

(* Note: this case was suggested as a good one for chapter three due to the topic of complaint handling. The same answer appears here if assigned as prescribed by the chapter)
Lynne should have begun by following procedure and not taken an order for MR Roast at that point of the evening. The kitchen should also have caught the error and reminded Lynne. She could have suggested another cut of beef that could be prepared MR. Failing that, Lynne should have informed Mrs. Wright of the doneness issue and let her decide if she would accept the beef (apparently she would have) or would like another cut. The host should have accepted Mrs. Wright's decision but may have thought she was simply being polite and wished to intervene on her behalf. The host may have made the error of being overbearing, but this may have been stimulated by the perception that Mrs. Wright was being disserved. The manager went overboard in removing all charges and offering free wine for the table. It probably would have been enough to offer to remove the charge for the roast beef dinner or provide a bottle of wine for the table, if they were drinking wine. Giving the table a new server may have allowed Lynne to be relieved from a situation that had gotten somewhat out of control and given the appearance of solving the problem. It also may have helped Lynne to "get back on the horse that threw her". As the table was now appeased, she probably could have resumed service. Lynne needs to be reminded of the procedures and the importance of guest service. She is a generally a good service, admitted her errors and went to the manager when the she saw the situation leaving her control. Retraining or simply coaching should do the trick. She may just have been having a tough night due to being tired. Firing her would surely be a mistake. A warning should be plenty as she seemed well aware of the consequences of her actions.

Chapter 5 Test

True/False

T F 1. A banquet can encompass any meal period and can be a special celebration or professional meeting. It may be preceded by a reception with beverages and hors d'oervres.

T F 2. Flying service refers to the English custom of serving guests from platters at a long table whereby the host carves the meat and the hostess passes plates to guests.

T F 3. Showroom service offer a meal and entertainment at a nominal fee in order to attract patrons to the facility where gambling often occurs.

T F 4. People who dine out to celebrate an occasion may want a bit more elaborate meal than they normally consume.

T F 5. An action station refers to counter service and all the functions that must be performed by the server to ensure speedy efficient service for guests in hurry.

Multiple Choice

6. The most common service styles for banquets are American and:
   a. French
   b. English
   c. Russian
   d. Self-Service

7. In setting up for banquets, it is common practice to:
   a. assign servers their stations and have them be responsible for each aspect of setting it up to promote accountability and mise en place
   b. have housekeepers from the hotel assist with the set-up due to their superior knowledge of linens and their ability to spot details servers may miss
   c. leave tables bare except for salt, pepper and candles so that guests can be brought exactly what they need and avoid feeling crowded by glasses and flatware
   d. preset all flatware china and glassware and sometimes, the first course, if cold

8. In banquets, a lineup or Preshift meeting may be used to:
   a. determine who gets to wait on the head table
   b. give specific instructions on setting up the unique function and review procedures
   c. review complaints from the previous function and threaten servers who fail to perform with unpopular closing duties
   d. inspect server readiness for the function to come and make sure everyone who was scheduled is on duty and on task

9. In the wave system:
   a. the entire banquet room becomes one large station where service personnel synchronize functions, starting with the head table
   b. departing guests are thanked for their business and asked sincerely to return soon
   c. sports fan are served in their stadium seats by roaming vendors
   d. each banquet or buffet server is assigned a station of 24-32 covers

10. In banquet table service using American style service all foods are:
   a. Served from trays with covers to preserve heat and allow for stacking of dishes
   b. Served from the right using the right hand and removed from the left using the left hand
   c. Served from the left side using the left hand and removed from the right using the right hand.
   d. Served whichever way is easiest at the time. Nobody really cares about which side anything is served from anymore nor do they notice.

11. Receptions are:
   a. always held in conjunction with wedding parties
   b. held after banquets
   c. used to fine tune microphones for banquet speakers
   d. gatherings to celebrate special occasions or to honor a person or persons

12. An American breakfast includes:
   a. Sugary cereal and caffeinated beverages
   b. Juice, cereal, entrée item, bread and hot beverage
   c. Fruit or juice, coffee or tea, bread or rolls and butter
   d. Fresh fruit, followed by a breakfast entrée and bread and beverage alcohol such as champagne.

13. A high tea can be described as:
   a. The most elaborate low tea but some heavier foods may be offered. The service is always seated and is usually not practical for large groups.
   b. Tea with sugar or honey, milk or cream and lemon slices on a tray
   c. A late night pick-me-up following the theatre and beloved by the British
   d. Served in silver pots with doilies placed under the saucers by butlers or maids

14. A complex buffet with a large offering of roasts, including game, rye break with sweet butter, caviar and vodka is:
   a. Hunt breakfast
   b. French
   c. Scandinavian
   d. Russian

15. What kind of service is provided on delivery carts that are brought to the bedroom of the ordering patron?
   a. Pizza
   b. Dormitory Cafeteria

c. Room Service
d. Fountain Service

Short Answer/Essay

16. What are the common table shapes for use with banquets? Why are they used?
17. What are the positives and negatives regarding using buffet service for large numbers of people?
18. Discuss the elements of providing kosher service.
19. What are the two or three functional stages of providing drive-through service? Why have some operations gone to three stages?
20. In reference to buffets and cafeterias, what is a shopping center arrangement?

Chapter 5 Test Answer Key

True/False

1. True
2. False. Flying service refers to servers presenting trays of beverages or appetizers on flying platters to guests who are standing. The Russians influenced the original French reception custom and introduced the concept of standing while eating.
3. True.
4. True
5. False. An action station is a buffet service concept that consists of cooking or finishing some food on medium sized pans over portable rechauds as guests go through the buffet line.

Multiple Choice

6. c
7. d
8. b
9. a.
10. c
11. d
12. b
13. a
14. d
15. c

Short Answer/Essay

16. Rounds seating 6-10 covers are the most common size used. Rounds are preferred as they are more suitable for conversation and service. Rectangular 6, 8 and10 foot tables are also common. Though they require more space, a larger number of patrons can be seated by using rectangular than with rounds. Square, oval, half rounds and serpentines are used less frequently. The table shape and capacities are chosen to best suit the function and as specified by the host/ess.
17. Buffet service is recommended for fairly small banquets as a large number of guests may end up milling around and creating confusion. It is difficult to move them along quickly. If the number of courses are limited, and shopping center approach employed, such as different entrée and dessert tables and salads preset, it may work.
18. Kosher foods need a Jewish person trained to oversee the complex preparation of food. Jewish dietary law strictly forbids that meats and all food prepared with meat products come in contact with dairy products. Separate kitchens or sinks, equipment and utensils must be provided to prevent mixing meats and dairy. All kosher food must be handled, prepared and served in full accordance with Jewish dietary law. Only the consumption of meat from animals that have split hooves and chew their cuds may be eaten, eliminating pork. Only certain birds, including chicken and turkey are allowed. Only fish that have both fins and scales are permitted. This excludes all shellfish.

19. Placing the order using two-way communication and paying and picking up the food. In a three stage process, paying and picking up are two distinct stages. This is done to enhance sanitation and speed service.

20. A shopping center arrangement is where different groups of foods are obtained from separate sections or areas. This is done to speed service.

Activity/Training

If the class is held in conjunction with a lab, it would be the ideal time to have scheduled a buffet for the complimentary service. At any rate, students could be broken into groups or individuals and assigned to design a menu and service for: A banquet luncheon honoring the retirement of their instructor, a reception celebrating their graduation from your institution, a banquet dinner for the homeless in your community, a buffet to thank the custodians and maintenance workers for the work that they do, a smorgasbord for anyone curious about Scandinavian cuisine, a cafeteria experience that is an alternative to what is currently provided at your institution, a high tea for administrative assistants, room service late night menu for a group staying in a suite, counter service for restaurant and hotel employees getting off work at midnight and fountain service for an area third grade class.

Chapter 6: Service Areas and Equipment

Objectives

- List and Describe the equipment typically found in the dining area of an operation as well as items used in table service.
- Describe the traditional hierarchy of a service staff

Outline

I. Introduction

II. Dining Area Equipment

A. Personal Items

- Servers should always carry an order pad or checks with a pen or pencil
- Serviettes or service napkins are used in very formal service
- Lighter or matches can be carried to light candles
- Are smoking sections growing to be a thing of the past? Servers used to light cigarettes and empty ash trays. (Legislation has gained momentum to do away with smoking in restaurants, bars and other public places and is diminishing the amount of smoking in restaurants and accompanying service that was provided for smokers)

B. Service Stations

- Also known as the Wait Station, the Service Station assists the service staff in performing their duties. They may be small with a minimal amount of dishes and flatware or large with many types of equipment.
- In a banquet operation, it may be as simple as a table along a wall. In a large, chain quick-service operation, it might take up 50% of the floor space
- All service stations perform one or more basic functions:

    1. Storage Center

        - The service station as storage center varies greatly between operations between simple and complex
        - Included may be; temperature storage units for hot and cold prepared foods
        - Cold storage includes: Ice machines, soda dispenses and milk machines, dessert cases, reach in refrigerators for pre-made salads, garnishes and side dishes.

- Warm storage includes: soup wells and bread warmers
- In each situation, server may need to plate or bowl the item to be served
- Just how much food preparation is it fair to have servers, often paid well below minimum wage to perform? Why are servers asked to plate up food? (Cost savings as servers make less than cooks, efficiency- so guest can quickly be served, and custom)
- Operators try to decrease the amount of space that doesn't have direct contact with the customer- therefore the space in service stations is decreasing- making vertical space expand
- How else might service stations be expanded? (Built into décor as with cabinets in dining room, more condiments etc. on table so less needs to be brought, mobile carts with service items etc.

2. Cleaning Area
    - It may include a place for dirty dishes, linen and garbage. Most will have sanitizing solution to clean bussed tables so they can be turned.
    - Items to aid in post shift cleaning- vacuums, polish and cleaners, are often stored
    - Strict sanitation rules must be enforced by management and are reinforced through certification training, required by law in many locales. Rules are consistent throughout the operation, holding temperatures, for example
    - Dirty utensils, dishes and linens need to be kept from clean items and food. Some states require a designated hand washing sink in kitchen and service areas that is used exclusively for that purpose.
    - If there is a lab associated with the class, students may be assigned to make a list of items stored there, sanitation procedures to be followed etc. If not, students may be asked to research requirements for their location and report back.
3. Command Center
    - POS systems allow servers to place orders to the kitchen, prompting needed information and recording information while tracking open checks, providing customer counts, and monitors the amount of sales
    - Bulletin or other announcement board may list soups and specials of the day, 86's, schedules, station assignments or other special information

4. Production Center

- Many operations now require service staff to be involved in some aspect of production such as; making coffee and tea or preparing an entrée salad.
- Beverage orders are commonly filled in the service station such as simple beverages, malts, shakes, premium coffee and specialty drinks which may involve bartending duties
- Production of food is generally limited to cold food that is prepped beforehand but does not hold well in the final state including; ice cream, desserts with whipped cream and salads. Servers assemble prepped ingredients to order
- Ask students for experiences in this regard. Is more food wasted in this fashion? Do employees tend to pilfer food that is available for their consumption? (various)

C. Host or Hostess Stations

- Area where the customers often first have contact with an employee
- Most have a host/ess stand, phone, reservation book, menus and a floor diagram
- Some have POS systems that enable them to function as a cashier

1. Kitchen or Service Area

- Manager or expediter is in charge of the area and acts as a communication link between the servers and kitchen staff and may clarify orders for the kitchen with servers
- Most operations that use an expediter require kitchen and service staff to communicate through that position.
- Expediters help servers with tray orders and assist in any production the server might have to perform
- What are the benefits of having an expediter position? (Centralizes communication with someone skilled at organization and prioritizing- helps maintain team atmosphere)

D. Furniture

1. Tables

- Tables for seated service are used in most operations the alternative being counters with stools or counters at which people simply stand
- Tables are sized to accommodate different sized groups and may be put together for larger groups

- Table arrangements are dictated by serving needs and the dining area shape
- Whatever the arrangement, it should be symmetrical and neat
- Levelers are used to keep tables from wobbling and should be used by servers
- Tables should be far enough apart to permit easy passage of people going by and not to allow a low conversation to be heard
- Have students ever experienced a "trough" like table arrangements whereby a row of deuces are only inches apart in places with cramped room or high rent- NYC for example? Describe experience. (Those who have, may have felt uncomfortable, privacy compromised, intrigued by others conversations, inhibited etc.) Inquire as to expectations of fine dining versus quick service in privacy factor. (Fine dining expected to allow more privacy due to nature of meal etc.)

2. Chairs
    - Sizes and shapes vary but all should be very sturdy
    - Usually, chairs do not have arms
    - Chairs should be large enough so guests can sit comfortably
    - Rollers make chairs easy to move but can be dangerous. Glides help chairs move easily over a surface
    - Chairs must be clean and in good repair, rough edges or legs can snag on guests clothing
    - High chairs or booster seats may be needed for children
    - Have students "shop" for restaurant chairs online etc. and see prices, styles available. You may also ask them to look at tables and other furniture.
3. Other Furniture
    - A gueridon is a small cart often carrying a rechaud, a small heater on which servers cook or reheat foods and liquid
    - Carts called voitures are used to move appetizers, desserts, wines, liqueurs or other categories of items around the dining areas. Some may be refrigerated or heated to hold items for carving and serving
    - Bus carts are used by servers or bussers to hold used wares. Most operations do not allow the carts to be brought into the dining room when guests are present

- Silencers are a felt pad that quiets the noise of dishes and other utensils and absorbs spills
- Tablecloths are commonly cotton, ramie, or linen and should fit and hang down only about eight inches
- Fine dining operations may use a naperone to cover the tablecloth. They are square and don't cover the entire table
- Napkins are 18-20 inches square and are used in fine dining for dinner. Smaller ones may be used for breakfast and lunch. Paper napkins are used in casual dining, for cocktails and receptions.
- A serviette is a hand napkin made of cloth and hung over the arm of the server to protect the hands while serving hot items and as a general serving towel
- Placemats are individual covers about 18-24" long x 12" wide. They are made of various materials and may be decorated with advertising, information, games or have the menu printed on them.
- Have students design an adult fine dining placemat. (Various- they can have fun with this) Alternatively, they could be asked to collect some placemats and bring them in to show the class.

III. Table Service Equipment

A. Dishes

- Dishware is expensive (plate may be $12 or more) and servers should give them good care
- Sliding dishware from a tray onto a dishwasher table should never be allowed. Flatware must be kept separate from dishes, sometime by using a special container
- Glassware should be given special care and kept separate from other dishes
- Heavy dish baskets or tote boxes (bus tubs) laden with soiled dishes must have a place to be set down. Areas where soiled items are brought should not be crowded and allow for cross-traffic
- Most dishware is heavy reinforced china (porcelain) in most operations. Lighter china is often too fragile and expensive. Ceramics, plastic and paper are also used. All must be sanitary and clean- with china sparkling
- Chipped, cracked or missing glaze are criteria for discarding glassware or china
- Most used dishes:
    1. Platter- long oval used to hold several portions or instead of plates. Sizzling platter are almost always made of metal and often use an underliner

2. Service/show plate- larger than the dinner plate, sitting at the table at the cover when guest is heated. First course foods in their serving dishes are placed on this plate. When removed, they are not used again for the meal. 11-12" and can be expensive- they are never brought to the dishwasher
3. Dinner Plate- used for dinner and lunch to hold the main part of the meal 10-11" and may be replaced by oval shapes instead of round
4. Salad Plate- 7-8" plate for salads, soups in bowls, desserts and other foods
5. Bread Plate- or bread and butter plate (B & B) 4-5" for bread, rolls, toast, butter, jams, jellies etc. Can be used as underliner with doily
6. Demitasse- round container with handle holding 4-6 ounces of hot beverage with saucer for espresso and specialty coffees
7. Finger bowl or monkey dish holds warm water with lemon or rose petal with underliner with doily.
8. Cocotte- small casserole for cooking and serving
9. Cloche- a bell shaped cover used with a serving dish to hold the heat in. Made of plastic, glass or metal
10. Snail dish- metal or ceramic dish with six deep indentations to hold snails cooked out of their shell, shallower indentations are used for snails cooked in the shell
11. Sauce boat- for serving gravies, au jus and other liquids

B. Glassware

- All glassware should be sparkling clean and sanitized
- Use of all purpose glasses saves money on inventory and makes life less complicated, specialty glassware can spark interest in beverage sales. Have students discuss the merits of each decision. Do any of them have a "souvenir" glass obtained from an establishment? (Various)

C. Service Utensils and Flatware

- All operations have special utensils used only be service staff- what those are depend on the service staff
- Most common are; ice scoops, tongs, ladles, stirrers and spatulas
- Other commonly used utensils include:
    - Dinner fork- for main entrees and other food eaten from a dinner plate
    - Salad fork- for salads appetizers, some desserts or fruits
    - Fish fork- for fish and sometimes seafood dishes
    - Cocktail fork- used for seafood and other cocktails
    - Lobster fork- when served in the shell
    - Dessert fork- pies, cakes, pastries and other solid desserts
    - Oyster fork- for clams, oysters and other bivalves

- Fondue fork- long for picking up bread cubes to dip into cheese fondue. A shorter fork holds meat in hot oil for meat fondue
- Teaspoon- for foods that are difficult to eat with a fork
- Tablespoon- larger than a teaspoon, used for soups or cereals
- Soup spoon- for soup
- Coffee spoon- for beverages, some cocktails and ice cream
- Espresso spoon- for liquids served in demitasse cup
- Sundae or iced tea spoon- for sundaes, and tall beverages served in deep glassware
- Sauce spoon- a wide shallow spoon used for sauces
- Snail tongs- for holding snails in the shell for extraction
- Lobster tongs- for holding lobster in the shell
- Pastry tongs- for picking up and serving
- Cake or pie server- for cakes, pies, pastries and similar desserts
- Why bother listing and describing these? (When servers are trained in what flatware is for, they can properly set the table correctly to make the guest experience as pleasant as possible. The right flatware provides the right tool for the job, making the guest dining experience as simple as possible)

D. Other Items

- Other commonly used items include: pepper mills, water pitchers, coffee servers, teapots, bottle openers, oil and vinegar cruets, mustard holders, bread baskets, cheese graters, service trays, tray stands, check holders and guest caddies.

IV. Menus

- Menus inform guests what your operation offers and may take the form of a traditional individual menu, sign, or board above the counter
- Some operations have servers recite offerings which limits the menu size but adds atmosphere
- Menus are usually printed on hard paper called Bristol or cover stock but could be on a sign board or placemat or even an order ticket
- A meal plan lists the type of food served at each course of the meal while the menu lists the exact items served (meal plan= soup, menu= bisque)
- Menus may be designed for specific meals or occasions; breakfast, lunch, afternoon, evening, room service, or banquet
- Menus may be designed for specific personal needs; children's, early bird, or vegetarian
- Some menus are not intended for guests but for the kitchen to inform cooks of preparation needs (production menus)
- Menus are sometimes named by their nature:
    - A la carte menu prices all items individually

- Table d'hote menu prices food together in a group often as a nearly complete meal
- Du Jour menu lists foods available only that day
- Cycle menu is one that runs for a period of time with foods changing daily, and then the cycle is repeated.
- California menu is one that lists items from all times of day all on one menu
- General menu is the main menu of a hospital from which special diets are planned

- Foods are customarily listed in the order for which they are eaten. Sometimes an operation has a separate menu for appetizers or desserts
- Menu items often carry some designator indicating how they are prepared
- Servers should know terms and find out if they are unsure
- Servers have to know preparation times so they can inform guests and plan their service time schedule
- Experienced servers develop intuition about time needed
- Beverages are listed in various ways depending on the kind of situation. Banquet or fine dining situations often list wine to the right of the menu item it is served with or beverages may be listed on the bottom
- Some operations have a separate menu listing all beverage alcohol
- Wine lists usually separate domestic wines from imported ones and may be numbered and coded on the menu
- Ask how class thinks types of menus are determined. (demand) What kind of food might be found on a board above a counter? On a piece of high quality Bristol? Hand printed on a blackboard at the door? Recited by the server? (quick service, fine dining, entrepreneurships of various description)
- Does the menu dictate service? Would breakfast service be different from a children's menu? How? (breakfast probably fast, efficient, and based on type of food, children's fast, friendly, casual)

V. Service Staff

- Two staffs are used; kitchen and servers. Servers are in frequent contact with the production type when they order and pick up. Knowing how the kitchen works can help servers operate

A. Kitchen Staff

- Cook usually directs other kitchen workers but frequently receives direction from an owner, manager or assistant manager
- Kitchen staffs can grow to a considerable size with specialized departments headed by someone directing the work
- Overseer of the commercial kitchen may be a kitchen manager or chef

- In health care facilities it may be the dietician
- In French organization there is often a continental kitchen with an executive chef in charge.
  - A steward orders, stores and may assist with menu planning
  - Sous chef runs the kitchen for the executive chef
  - Chefs de parti include; garde manger (cold foods and pantry), patissiere (bake shop), legumier (cooked vegetables, garnishes), potager (soups and stews), entremetier (sauced and roasted dishes) assisted by helpers
  - This system was introduced by Chef Auguste Escoffier
  - The aboyeur announces the orders from the server to the cooks (much like an expediter)
  - Checkers make sure orders are correct, price them and give them to the aboyeur to call out.
  - Other helpers include; sanitarians, and other employees
  - Friendly relationships can be very helpful in productivity when everyone is under stress
  - When there is a rush, it is wise not to make special demands
  - Sharing tips with cooks can create a high degree of cooperation between cooks and servers
  - Discuss the all too often conflict between the front and back of the house. How does pay play a part in the conflict? (Cooks are generally hourly or on salary and may receive bonuses for achieving food cost goals. Servers often have a greatly reduced hourly rate and their pay is largely provided in the form of gratuities from the guests. The servers income may depend on receiving quality, correct product from the kitchen who often have no direct financial incentive to get it right for the guest)

B. Traditional Service Staff

- Service staffs vary considerably
- Management hires, makes schedules, assigns stations, sets up service standards and training and directs other service work
- Some operations such as drive-in and quick service have little or no staff organization
- Seated service operations are more complex. Manager is in charge with one or several servers working on a shift, if the staff is large, the host/ess may be in charge
- Most complex staff is European model based on staffs of Ritz and Escoffier. It is labor intensive and expensive but elegant and lavish
  - Manager of serving is the Maitre d' hotel, host or head waiter. They assign stations and may hire serving staff
  - Captains are in charge of a group of servers or chefs du rang or front servers. Each chef du rang is assigned a station of up to 25 seats.

- A Back server, assistant waiter or commis assists the chef du rang.
- Buspersons are called commis debarasseur
- In some French server staffing, the chef de service acts as maitre d' and may have an assistant, chef d' etage who directs serving in he dining room
- A maitre d' hotel de care supervises a dining room section, somewhat as captains do
- Food and Beverage manager is in charge of all food production and service,
- Banquet managers direct banquet staff
- Sommelier or wine steward is responsible for wines and other beverage alcohol. Some units use bartenders who only fill orders for service staff and not directly, the public
- Buspersons bring trays of orders to a station, take away soiled items and assist servers
- In healthcare, a dietician is in charge of service but nurses often deliver food sent up from the kitchen. Servers may serve in visitors or doctors dining rooms
- Have students look at various menus and propose a system of service organization suited to that type of menu (various)

Chapter Review Questions Answer Key

1. A service station assists the service staff in performing their duties. It is normally used for storing basic goods, holding items for cleaning the dining room, acting as a command and communication center, serving as a production center, possibly may serve as a host/ess station and acts as a link between the servers and kitchen staff. It normally holds dishes, flatware, table condiments, various paper goods, hot and cold prepared foods, glasses, cups, linens, trays, and tray stands and may also hold equipment as well.
2. Matching terms and definitions:
   (1) Gueridon (e) used to hold food and equipment for tableside preparation
   (2) Naperone (c) covers tablecloth
   (3) Snail tongs (b) used with a dish having small indentations in it
   (4) Demitasse (d) holds about 4 ounces of espresso
   (5) Cocotte (a) small casserole
3. Matching terms and definitions:
   (1) Aboyeur (g) calls out food orders
   (2) Maitre d' (c) in charge of a group of waiters
   (3) Commis debarasseur (d) busperson
   (4) Commis (a) assistant waiter
   (5) Chef d'etage (f) assistant to chef de service
   (6) Sommelier (e) in charge of wine service
   (7) Chef de rang (b) front waiter
4. A cycle menu is one that runs for a period of time with foods changing daily, and then the cycle is repeated.

5. Items are presented on the menu generally in the order that they are eaten, that is appetizers, soups, salads, entrees and desserts. There may be separate menus for appetizers or desserts. Less usual, menus may list different food groups by courses.
6. (Production) menus are not designed to be used by guests for ordering but to inform cooks of preparation needs.
7. Cover stock is called Bristol and refers to the hard paper menus are printed on
8. A gueridon is a small cart often carrying a rechaud which is a heater on which foods are cooked or reheated. Voitures are carts used to move hors d'oervres, desserts, wines or beverages (or other courses) around the dining area. Offerings should be neatly arranged. Carving may be performed from the cart but cooking is not.
9. A la carte prices all menu items individually and table d'hote prices food together in a group often as a nearly complete meal.
10. Answers will vary. The advantages include creating a high degree of cooperation often not found in a historically adversarial relationship. As servers often suffer by being tipped less when the kitchen makes an error outside of the servers control, it may increase the cooks awareness of the ultimate goal of pleasing the guest. It may work if cooks hourly pay is somewhat reduced and server pay increased, otherwise servers may perceive simply another hand out to take another piece of a pie already dwindled through tipping out buspersons, bartenders, cocktail servers and paying taxes. Another advantage is that server pay is not as dependent on customer whim alone if hourly pay goes up. The disadvantage, (besides pay issue just mentioned) is that the guest is tipping the server and does not realize a crowd is involved.

Chapter Cases Answer Key

Server Stations

Various- look for the elements described in the chapter. It might be useful to have students critique the designs. For privacy purposes, names can be kept off the work that others will critique.

Chapter 6 Test

True/False

T F 1. Sanitation rules applying to the kitchen and dining areas are different due to the different nature of activities, light levels and sensitivity to chemical smells experienced by many guests.

T F 2. Today's multi-tasking servers are often required to plate cold food items and prepare hot food for their guests from the service station.

T F 3. The host/ess station is most often the area where customers first have contact with an employee.

T F 4. The closer together tables are the more of a sense of cozy hospitality experienced by the guest. Diners generally prefer this arrangement.

T F 5. Usually, dining room chairs do not have arms.

Multiple Choice

6. In very formal service, a server generally carries:
   a. A bus tub to clear dirty tables quickly
   b. An order pad and writing instrument
   c. An order pad or checks, pen, serviette and matches or a lighter
   d. A wine opener, cash bank and decanter

7. A silencer is:
   a. used to cover the tablecloth in fine dining
   b. engaged to muffle the sounds of guests who are enjoying themselves too noisily
   c. a buffer zone constructed to keep the noise from the kitchen from disturbing the guests
   d. a felt pad that quiets the noises of dishes or utensils and absorbs spills

8. Dinner napkins are generally:
   a. 18-20" square and made from cotton or linen
   b. the same size as breakfast or lunch napkins
   c. cleverly folded and made from recycled, unbleached paper
   d. found folded over the arm of the wine steward and used to prevent the cork of red wines from making an embarrassing popping sound when removed
9. A serviette or hand napkin is usually made of cloth and serves the purpose of:
   a. assisting guests in cleaning their fingers following a messy dish such as barbeque ribs or corn on the cob
   b. protecting the hands of the server when serving hot items and as a general serving towel
   c. absorbing condensation from a cocktail glass in the bar or at the guests table
   d. polishing the spots on glasses and flatware caused by hard water

10. The following statement about placemats is NOT accurate:
   a. They are individual covers about 18-24" x 12"

b. They may include games for children
c. They are used in fine dining to provide an additional layer between the table and tablecloth providing sound and spill absorption
d. Some have the menu printed on them

11. A dish used to hold several portions or as an alternative the traditional shape that may be metal so that it can be heated to a high temperature so the food sizzles on them as they are taken to the guest is called a:
    a. Show plate
    b. Cloche
    c. Snail dish
    d. Platter

12. A cuplike container holding lukewarm water with a slice of lemon or rose petal floating in it is called a:
    a. Teapot
    b. Finger bowl or monkey dish
    c. Sauceboat
    d. Demitasse

13. A menu that prices food together in a group that is often a nearly complete meal is:
    a. Early bird
    b. General Menu
    c. Table d' hote
    d. A la carte

14. In the most complex service staff system, modeled after Ritz and Escoffier, the manager of serving is usually called a:
    a. Captain
    b. Commis
    c. Maitre d' hotel, host or headwaiter
    d. Dietician

15. The person who works with the chef in ordering, storage, perhaps menu planning and is in charge of much of the dining room equipment and napery is the:
    a. Steward
    b. Sous chef
    c. Chef de rang
    d. Garde Manger

Short Answer/Essay

16. Describe, briefly the basic functions that may be performed by the service station.

17. Explain some of the features of a point of sale system as it relates to the service area function, "command center".

18. Describe the function of an expediter in kitchen as the position relates to service.

19. Explain the difference between a gueridon and a voiture.

20. What is the purpose of a service or show plate?

Chapter 6 Test Answer Key

True/False

1. False. Rules applying to kitchen and dining areas should be the same as the requirement is to serve safe food.
2. False. Servers may put together cold items from prepared foods or plate (or bowl) hot food but hot food is seldom if every prepared in a service area.
3. True.
4. False. Tables should be far enough apart to permit easy passage of people going by and to not allow a low conversation to be heard at an adjoining table.
5. True

Multiple Choice

6. c
7. d
8. a
9. b
10. c
11. d
12. b
13. c
14. c
15. a

Short Answer/Essay

16. Service stations store basic goods, hold items for cleaning, act as command and communication centers, serve as production centers, sometimes serve as host/ess stations, and provide a link between the servers and kitchen staff.
17. The point of sale, or POS system allows servers to place orders to the kitchen in an orderly and traceable way. It records and informs the kitchen of the customers order, prompts needed information, tracks open checks, keeps a customer count, totals check sales and monitors sales amounts.
18. The expediter is in charge of the communication between the kitchen and the servers. They relay messages to cooks from servers and clarify server orders for cooks. Most operations require that the expediter be used as the sole means of order communication. They may help servers' tray order and assist with server production tasks. The expediter may be used instead of or in addition to a point of sale system.
19. Both are carts which are used by servers to transport food but the gueridon has a rechaud to reheat or cook food or liquids, the voiture is more often a display of a category of food of drink. The voiture may contain chilling or warming units and carving or serving may be done from them but actual cooking is not performed.
20. The service or show plate is used in fine dining to sit on the table at the cover when the guest is seated. The first course is served, and the serving dishes are placed on this plate.

Once removed, show plates are not used again. They are 11-12 inches in diameter, expensive and require careful handling. They are generally wiped clean with vinegar and never brought to the dishwasher.

Activity/Training

In situations where a lab is a part of the class, the service stations may be organized, inventoried and cleaned by class members. Further, the service system may be re-engineered with different ways of providing service tried- four person teams, busperson/server teams, or lone servers with smaller stations and more duties.

Classes which are not accompanied by a lab might be given a variety of menus from which groups may devise service systems, and develop an arrangement of service stations deemed useful.

Chapter 7: Classic Service Styles

Objective

- Describe the techniques of French, Russian, American, English and Chinese Service Styles

Outline

I. Introduction

- The different ways to accomplish serving food and beverages is referred to as service styles

- The service style used by a facility should not go by a rigid definition but should be used to best suit the facility served. Authorities do not always agree on the details and therefore, it is possible those in the chapter could be challenged.

- Service style is the way food is presented to the guest and service concept is the type of service offered.

- There are as many styles as there are cultures. This chapter presents the five most traditional or "classic" styles

    - French or cart or gueridon service

    - American or plate service

    - Russian or platter service

    - English or butler service

    - Chinese or family-style service

II. French Service

- French service evolved from the style of the French Court, le grand couvert (the great cover) and has been considerably modified

- The distinctive feature is the preparation of food is at tableside and is served from the service table to the guests

- Final food preparation is done where guests see it and is a show requiring skill and finesse to do properly

- It is slowest, most labor intensive, slowest and most expensive

- Guest turnover is reduced, ware washing costs are increased. More space is required so fewer tables can fit in the dining room. The high cost of such service draws only a few diners willing to pay

- The gueridon is the cart used and the heater used to cook is the rechaud. The cart has the tools and equipment for cooking, tossing salads and deboning fish or carving meat

- Service staff consists of:

    - Chef de rang- main or front server- prepares, dishes and garnishes, food- may take and bring alcoholic (except wine) beverage orders

    - Commis de rang- assistant server-hands warm plate with food to guest

    - Commis de barasseur- busperson- assists chef and commis de rang and clears

    - Captain- direct supervisor of team- may perform more difficult preparations, deboned or carve

    - Maitre d' hotel or dining room supervisor may also assist with difficult preparations

    - Cocktail servers- serve alcoholic beverages except for wine (if no cocktail servers, may use chef de rang to perform this function)

    - Sommelier- wine steward- performs wine service

    - How might this complicated system be modified for fine dining in situations where there is little if any tableside service? (various- it is sometimes done with plated food whereby there is a captain who takes the order and communicates most directly with guests and may also perform cocktail and wine service. The front waiter serves the food at the table, the backwaiter brings the tray of plated food to the table for the front waiter to serve, the busperson clears dirty dishware, refills nonalcoholic beverages, brings rolls, butter and condiments such as sugar, etc.)

    - Typical place settings vary but limit the amount of flatware and glasses with more being brought out as needed. Salt and pepper are rarely on the table and bread and butter plates are not always used. No ashtrays are on the table. If smoking is permitted by law, it is "properly" done after the meal.

- Why are salt and pepper not typically on the table in French service? (It is assumed that the food has been properly seasoned. Americans, who are accused of over salting their food or seasoning food prior to tasting it are sometimes put off by this custom)

- Is the customer always right? (No- but the guest is always the guest and should be treated as such.)

- The meal usually ends with finger bowl service

A. Tossing and Mixing

- Used primarily to make salads and should be done working toward to the server to avoid the possibility of spattering the guest

- Mix silently and gently using quick, short motions

- Add oil first so it coats the salad and then vinegar, reversing the order may cause the oil to run uselessly to the bottom of the bowl. Caesar salad is the exception where coddled egg is added first which tends to absorb rather than repel oil

- What other sorts of salad lend themselves to tableside presentation? (Classic cobb salad and other chopped salads may be tossed easily at tableside so that presentation is enhanced)

B. Plate Presentation and Sauces

- Food should be placed in the center of the serving dish so the rim can act as a frame surrounding the food.

- Sauces are common in French food and be presented under the entrée or poured over it, or on the side in a sauceboat or gooseneck. Lemon served with fish may be wrapped in cheesecloth to "strain" seeds out

C. Deboning and Carving

- Experience and knowledge of the bone and muscle structure of proteins help.

    - Fish have bony fins and a skeletal structure in which the belly bones are attached to the spinal bone
    - Poultry has a keel or breast bone and legs and wings attached to the backbone
    - Bone and muscle structure of meat will depend on the cut but many will have much of their bone structure removed prior to cooking.

- Deboning Fish

    - It is important to leave the deboned parts whole

    - First, remove fin bones from the middle of the back, belly and near tail. Place a fork into the fish and using a spoon's edge, pull the fin bones away from the fish.

    - Next, with the fork, run down the back under the flesh and lift up. The top fillet should be easily lifted up as a whole piece which is laid on the serving plate

    - Carefully remove the meat covering the belly bones and lower tail

    - Next, by lifting the fork at the tail, remove the spinal column from the other half of the fish, with it should come the belly bones

    - It is important to see that all bones, even tiny ones are removed

    - Having class deboned fish would be ideal but probably not practical, doing a demonstration or showing a demonstration video helps students visualize this technique.

- Tools include proper knife and utensils. A 10" carving knife is needed to slice roasts and cut through a rack of lamb and a 7" narrow, rigid knife is needed for poultry. Always use a sanitized cutting board

    - Most meats are sliced across the grain

    - Poultry such as duckling is split in two, laying the breast side down and using a sharp butcher's knife cutting through the back and breast

    - Keel bone is removed and ½ is served as a portion.

    - Chicken is not usually deboned entirely:

        - Wings are removed as is the wing tip, leaving two joints;
        - Legs are removed by pulling away from body and separating at joint. Thigh and leg may be separated in small birds.
        - By cutting down closely along ridge of keel bones, the two breasts can be removed whole.
        - Slicing can then occur.
        - In larger birds such as turkey or goose, no deboning is done but meat is sliced from bird on platter breast side up.
        - Practice, demonstration or video graphically assist students in understanding these concepts

- Chef de trancheurs are used in Europe to carve meat according to guest selection

D. Flambéing

- Most often done in French service, this method is not usually done tableside due to danger of accidents

- If done properly, it can be spectacular and entertaining

- No bottles of alcohol should be allowed to come near a flame because the gases can ignite, causing the bottle to explode and burst into flames causing death or disfigurement of the inexperienced server

- Flambéing does not involve cooking but warming. The flame should be kept low and never creep up the sides of the pan

- Rechaud and blazer or sauté pan are used. Shallow copper pans for crepes are sometimes used for that purpose

- Too much sauce overwhelms alcohol and may fail to ignite or burn poorly. There should be a thin film of alcohol covering the surface which is warmed to form vapors

- Vapors are ignited by tilting the dish slightly toward the flame

- A towel or serviette should be handy to quickly throw over the flame should any accident occur

- Wait for flames to die before serving.

- Common desserts include; Crepes Suzette, Bananas Foster, Baked Alaska and Cherries Jubilee

- Again, practice, demonstration or video are helpful. Though dangerous, this is the least expensive demonstration to have and students thoroughly enjoy the finished product

E. Fire Plan

- A plan should be established and every employee trained to protect and evacuate guests.

    - Call for the fire department or building protection unit from the burning property or outside if evacuation was rapid.
    - See what steps can be taken to put out controllable fire- smother with flame retardant draperies

- o Know where extinguisher/materials are. They should be nearby and accessible
  - o Turn off all possible sources of flame such as a gas line
  - o If evacuated, see that everyone is accounted for, safe and away from harm
  - o Follow any instruction of the plan

- Review plan for the dining room associated with the class or building the class is held in. Ask class for experiences with fire and for the cause(s). (Various)

F. Serving Hot Items

- Patrons should always be warned by servers if the food or plate might burn

- Steaks on sizzling metal platters and hot soups can cause burns. Hot tea, coffee and other liquids can be spilled by servers. Patrons must not hold containers while they are being refilled. Containers should not be refilled to the rim.

- Ask students for experience with serving or being served hot items (*Note, this author suffered serious burns on her chest when a guest rose under a tray full of sizzling prime rib platters with au jus. The platters and jus fell on the server's chest. Ice water was poured over this server who completed her shift. Remind class that dangers are present for servers as well as guests and to be mindful.)

III. Russian Service

- Russian service was greatly modified from the original French, becoming a unique style.

- All food preparation is done in the kitchen

- It is less labor intensive because only one server is used, is faster and requires less space than French

- Russian service is well suited to seated service of large groups eating the same meal and can be quite elegant

- Guests feel they are getting personal service because of the way they are served

- Ask students if they have experienced this type of service and where it might be utilized. (Cruise ships sometimes use this method, banquets or private parties)

- Dishing is all that is done at the table. All preparation and finishing has been done in the kitchen

- Liquids such as soups or sauces are placed in tureens and brought to the table with soup sometimes brought in individual serving bowls

- Food items are placed from a platter onto a plate directly in front of the guest from the left side

- Food is so hot servers may wrap their arm with a small towel under their jacket sleeve

- Server faces the guest and rests the platter on the left arm and left hand palm. Smaller dishes rest only on the palm requiring skill to balance and serve.

- Portions of food are lifted by placing a large spoon, bowl side up under a portion and a fork on top to hold. Spoons are used for vegetables and others

- Table setting is much like French but with an ashtray, water glasses, and bread and butter plates. Rules are not as formal or strict as with French service

- Servers move around the table clockwise, placing the dishes in front of guests from the right side using their right hands until all have their dishes

- Coffee is served at the end of the meal with cups and saucers brought and coffee then poured. Cream and sugar are brought if desired. Finger bowls are not usually served with Russian service.

IV. American Service

- American service is the fastest, least labor intensive, most commonly used and most adaptable to casual to formal dining

- Food is dished on plates in the kitchen and taken to guests, saving time and assuring temperature

- Fewer dishes are used, reducing usage and ware washing costs

- Food can be very attractive as it is plated and garnished by the chef in the kitchen in a visually pleasing way

- Space is minimized for service

- French and, to a lesser extent, Russian service are difficult to adapt to various meal periods and needs

- Standard American table setting is the most widely used table setting in food service

- Spoons and knife, with blade facing inward, are on the right and forks on the left with tips of handles about a ½" from the table edge. Flatware is normally placed according to the progression of courses working from out to in

- The amount of flatware used varies according to the menu but the limit should be three to either side with more brought as/if needed

- Service plate may or may not be used with hors d'oervres plate in its place. Bread and butter plate is above forks and sometimes slightly to the left with the butter knife at a right angle or parallel to the forks

- Water glass is to the far right above spoons and dinner knife with wine glasses to the left of the water

- Decorations and ashtrays, salt and pepper, and sugar bowls are typically in the center of the table or against the wall for booths

- Coffee cup and saucer may be on the table just to the right and slightly abode the knife and spoons with service during the meal or served at the end and/or brought just before being served

- With the popularity of coffee, will American service feature more cups pre-set? (maybe) Would the kind of restaurant or meal period effect whether or not coffee cups are preset? (Of course. A diner at breakfast would surely have coffee cups on the table while a fine dining dinner house may not bring cups until and unless coffee is ordered.)

- The table may or may not be covered with a tablecloth. Napkins can be folded in various ways and usually put on the center of the cover, in front of the guest

- The class may engage in napkin folding whether or not the class is associated with a lab. There are books, videos and websites to support this activity. Booklets are put out by most linen companies. www.wholesale-table-linens.com, http://www.napkinfoldingguide.com/, http://www.napkinfolding.net/, are but a few online guides to folding napkins.

- Service moves around the table from right to left, serving from the left, using the left hand and serving beverages from the right using the right hand. Rules should be flexible and keeping what is best for the guest in mind.

- Bread is normally served in a basket unless the setting is very formal and is served by the server using tongs

- It is not considered proper to remove any dishes or start a new course until everyone has finished, though this is frequently violated in busy banquet settings. Items removed from the right are done using the right hand and the left, the left hand
- Crumbing is removing all particles of food left on the table after the entrée in order to have a clean table for dessert and may done in more formal meals

- 15-25 covers are considered a good work load using American service. The team approach can work well using a modification of the French team mentioned earlier

- Ask students which service conventions they notice and which do not matter; serving females first, not removing dirty dishes until everyone finishes, serving from left, etc. (various depending on ages, backgrounds and preferences of students)

- If the class is associated with a dining room, comment cards should address service technique so that students may see how many guests do notice things that do not matter to the students. Should service be geared toward the market served? (Of course)

V. English Service

- Also called butler service, English service uses less labor than Russian and is a little bit faster

- English service is rarely used in the United States but is seen in some clubs and private parties.

- Would this service be suitable for a bed and breakfast? (Yes, or a family holiday gathering)

- Often the first course is on the table when guests enter with ice water poured and bread and butter out or brought as needed

- Soup may be served dished into individual bowls by the server or placed in a tureen in front of the host/ess who then serves, handing servings to be passed down or to a server who passes

- Meat or main entrée is placed in front of host with warm plates. The host serves a portion of the meat or entrée and passes it to the right or first carving and then passing, the hostess adds vegetables or side dishes and passes or gives to service personnel to pass.

- The hostess gets the first plate (the coldest) and the host the last. No one eats until the hostess lifts her fork and touches the food on her plate

- Table settings are the standard knife and spoons on the right and forks on the left with dessert flatware above the service plate, if one is used

- Desserts are plated and passed by the hostess who also performs coffee service with cream and sugar passed

- What are the class members' views on the usefulness of this style? Could it be modified for use in family style restaurants? (Various- it might be imagined as a version of the all-you-can eat buffet where food is brought to the table and the parents serve their families from platters. It could be said that this is the service when food is taken away and eaten at home)

VI. Chinese Service

- The setting for Chinese service is quite traditional originating with Chinese families and formalized for special occasions
- There are many customs that must be followed in order not to give offense
- If classmates seem puzzled or amused by customs, remind them that many mores of American service may seem just as odd to those not familiar with it.
- Labor is minimal but ware usage is extensive since foods are usually served in individual dishes
- Chopsticks are used for almost all foods except for soup and dessert spoons
- Food is dished into large platters, bowls and other containers and brought to the table where guests select their own food.
- A circular wheel (lazy Susan) may be used by Chinese food operations to help guests rotate and share food
- Normally a 6" small plate is placed in the center of the cover with a bowl of rice to the right and a cup for tea above the rice bowl
- Other dishes and utensils are set as the menu dictates
- Small dishes are provided for sauces and condiments and a soup cup with a porcelain soup spoon may sit at some location near the guest
- Food is brought to the table on platters or serving dishes and placed in the center of the table where guests help themselves to what they want from elaborately garnished dishes
- The order of food served including sweets, soups and rice may not be what Americans are accustomed to
- Large round tables are used with as many as 12 people
- The custom is for the host/ess to sit with their back to the door while the guests face it
- Usually no tablecloths are used, though napkins are

- All dishes except wine cups should be removed when they become empty.

- Chopsticks may be challenging so servers should be alert to the need to be provided a fork. Knives are not needed as food is cut into bite sized pieces

- Do students feel any other cuisine may lend itself to modified Chinese service? (probably anything that can be eaten family style)

Chapter Review Questions Answer Key

1. The five classic styles used to serve food and beverages are; French, Russian, American, English and Chinese

2. The distinguishing characteristics of the five classic service styles are as follows:

   French- food is tossed, heated, or prepared tableside by a skilled team of servers. This is the most labor intensive and expensive form of service and the most elegant.

   Russian- food is served to individual diners from a platter on the arm of the server. All food is prepared in the kitchen; it is a modification of the French style that is less labor intensive, elegant and suited to large seated groups eating the same meal.

   American is the fastest, least labor intensive style and most commonly used. Food is plated in the kitchen and brought to guests. It can vary from formal to casual service and requires minimal space.

   English or Butler service uses less labor than Russian and is a bit faster. It is rarely used in the United States but has graciousness desired by intimate groups.

   Chinese is quite traditional, originating with families and formalized for special occasions. There are many customary manners that are followed. Chopsticks are used for nearly all foods except soup and dessert. Rice is served with every meal and food is served on platters that are usually elaborately garnished. The diner eats soup with a porcelain spoon.

3. French service is elegant and the most formal, special kind of service. It is also labor intensive, time consuming and therefore expensive. Russian service is less labor intensive, faster than French and requires less space. Guests feel they are getting personal service because of the way they are served. It still requires skill and takes more time than American Service. American Service is the simplest and most versatile. Plated foods are garnished and plated by chefs and can be very attractive. It may have the least disadvantages as it is by far the one most often used. It may be said to be the most common and therefore least "special". English Service uses less labor than Russian and is faster. It is suitable for intimate groups. It is seldom seen outside of private clubs or parties and may have little practical application other than bed and breakfast type settings. Chinese service provides a good variety of food to which guests help themselves. The traditional customs are not always familiar to Westerners and may have limited application other than, of course in Chinese restaurants or modified to family style serving situations with other ethnic cuisines.

4. French Service matching definitions:
   (1) Chef de trancheur (c) Carves meat

(2) Chef de rang (d) prepares food at tableside
(3) Gueridon (a) a cart on which food is prepared at tableside
(4) Rechaud (e) used to heat food
(5) Sommelier (f) wine steward
(6) Flambé (b) to ignite
(7) Captain (g) head of service

5. In Russian service, soup is served to guests in large tureens, brought to the table or it is brought to the table in individual bowels.
6. In Russian service the server places a spoon under a portion of food, a fork over it and it is placed on the guest's plate.
7. A porcelain soup spoon is used in Chinese service.
8. Coffee cups and saucers are often on the table in American service at the start of the meal unless the service is for fine dining.
9. French service descended from the le grand couvert (the great cover).
10. When deboning a fish, the fins are removed first. There is one in the middle of the back, several on the belly and one near the tail. Deboning is a traditional part of French service.
11. Matching definitions of service styles:
    (1) Butler service (d) English
    (2) Food prepared in kitchen and served to guests by servers from platters and bowls (b) Russian
    (3) Food is dished onto plates in the kitchen (c) American
    (4) Salad mixing and meat carving tableside (a) French
    (5) Chefs have the most opportunity to garnish and plate attractively (c) American
    (6) Team serves guests (a) French
    (7) Ashtrays, sugar, cream, salt and pepper are on the table prior to guests being seated (c) American
    (8) Many foods are served by the server using two spoons (b) Russian
    (9) Food is sometimes covered with alcohol and set ablaze in front of guests (a) French
    (10) Server rests serving platter on the arm or open palm (b) Russian

Chapter Case Answer Key

Setting Up a Service Staff

The maitre d' will need service teams for the classical French style service. Assuming each team can handle 25 diners per evening, four teams will need to be on duty each night. As the restaurant is open seven nights per week, there will be nights that each employee will have off. Some employees may not need or want to be full time. Making these assumptions, the maitre d' will need 6 chef de rangs or front servers, 6 commis de rangs or assistant servers, 6 commis de barasseur or buspersons and 1-2 sommeliers. It may also be necessary to have 2 bartenders and 3-4 cocktail servers for the Las Vegas setting. There will need to be 6 captains of service, assuming there is one captain per service team, less if captains can handle more than one team. Training will be an important element in making the operation a success as this is not a style of service that is so common that many will have a thorough knowledge of it. Tapping into culinary and hospitality programs for employees may be a wise move.

Chapter 7 Test

True/False

T F 1. French service is also referred to as gueridon or cart service due to the tableside food preparation, mixing, deboning and carving performed in show style fashion for the guest by the server.

T F 2. French service evolved from the elegant and lavish style of the French court and retains a rigid adherence to the rules of service developed then, for royalty.

T F 3. A sommelier serves wine and other alcoholic beverages in French service

T F 4. Fingerbowls are typically used with Chinese service, due to the challenging nature of chopstick usage.

T F 5. When tossing a salad tableside, oil is added last so that other ingredients do not slide off of the salad pieces.

Multiple Choice

6. The term "plate" service is more commonly known as:
   a. French
   b. Russian
   c. American
   d. English

7. The term "family style" service is also known as:
   a. French
   b. Russian
   c. American
   d. Chinese

8. French service is the most expensive style due to:
   a. The cost of the fine food and wine
   b. The cost of the equipment needed- gueridons must be imported and the pots and pans are also very expensive
   c. Labor intensive nature of tableside service, slowness and ware washing demands
   d. Insurance for the inevitable flambé fire

9. Salad mixing tableside should be:
   a. Done by tossing toward the server using quick, short motions.
   b. Performed by mixing toward the guests with gusto to emphasize the showmanship of the server and the excitement of tableside service
   c. Accomplished by moving the bowl in a vigorous rolling fashion, much like sautéing vegetables, causing the ingredients to appealingly "jump" up from the mixing bowl
   d. Done in such a manner that everyone gets a little of each ingredient in their portion and there is nothing left in the mixing bowl

10. Which of the following is NOT a technique for serving hot items?
    a. Warning guests if the food or plate might burn them
    b. Taking care when serving coffee or tea, by not allowing patron to hold the container they are filling
    c. Leaving room at the rim of hot beverage containers so that they are not overfull
    d. Blowing on them to cool them, or, in extreme cases, popping them in the refrigerator to chill them to consumer friendly temperatures

11. Which of the following is a disadvantage of American service?
    a. It is the fastest form of service
    b. It offers the least amount of customization of the five classic styles
    c. It is the least labor intensive
    d. Food is plated and garnished by the chef

12. What is the average number of covers for a server using American service? What is the maximum number for dinner service?
    a. 8 and 16
    b. 15 and 25
    c. 20 and 30
    d. 30 and 50

13. What is the advantage of English service?
    a. It has a graciousness desirable with more intimate groups
    b. It allows the host to do some of the work, saving labor for the operation
    c. It is suitable for large banquets where groups of people are eating the same food
    d. Servers are referred to as "butlers", wear cummerbunds and are generally more respected than their counterparts in less formal service styles

14. Chinese service is referred to as family style. This is because of all of the following reasons EXCEPT:
    a. It originated with Chinese families
    b. Food is served from platters where guests help themselves.
    c. Large round tables for up to twelve people are used.
    d. The economical advantages of sharing food for family groups.

15. Which of the following is NOT a feature of the Chinese style place setting?
    a. Soup is eaten with porcelain spoons
    b. Chopsticks often rest on a small stand
    c. Sharp knives are provided
    d. Small dishes for the sauces and condiments are set at each place

Short Answer/Essay

16. Discuss what skills, knowledge and tools the server should possess in order to debone foods well.

17. Describe the process of flambéing safely. Why do guests like to order flambéed desserts?
18. Explain the difference between French and Russian service. How are they related?
19. Describe how food is served to the guest using American service. Be specific about technique.
20. Explain the reasons for the overwhelming popularity of the American service style.

Chapter 7 Test Answer Key

True/False

1. True
2. False. French service has been considerably modified and many variations have been introduced.
3. False. The sommelier does serve wine but does not serve other alcoholic beverages. This is done by the chef de rang or cocktail server.
4. False. Fingerbowls are used most typically with French service, though they are sometimes offered with Russian and even formal American service.
5. False. Oil is added first so that it does not end up sliding into the bottom of the mixing bowl. The exception is Caesar salad where it is added after the coddled egg which has the capability of absorbing the oil.

Multiple Choice

6. c
7. d
8. c
9. a
10. d
11. b
12. b
13. a.
14. d
15. c

Short Answer/Essay

16. It takes experience to debone foods well. The server needs to know the bone and muscle structure of each item. It is important when deboning fish to leave the deboned parts whole. It is critical to use the correct knife and proper utensils. The cutting board should always be sanitized prior to use.
17. The bottle of alcohol should not come too near the flame. The amount of alcohol should be enough to ignite, just covering the surface with a thin film. The alcohol enriched sauce should be warmed so that the alcohol vapors are formed. The flambé pan or dish should be tilted slightly toward the flame, which should ignite and spread rapidly over the surface. A towel or serviette should be ready to smother a wayward flame, depriving it of oxygen. When the flames have died out, the item can be served. Guests like to order flambéed desserts because the presentation is spectacular and entertaining (and it tastes good!)
18. French service involves a team of servers, with defined, hierarchal roles who mix, prepare, debone, carve or flambé food at the table of the guest, before their eyes. It is elegant and requires a high level of skill. It is the slowest and most expensive type of service. Russian service requires less space than French service. The food is prepared in the kitchen and placed on platters or other dishes. These are brought to the table by a single server who balances the platter on arm or palm and uses two utensils to dish food onto the guest's plates. It is well suited to large groups eating the same meal and provides the feel of personal service to the guests. Russian service originally came from French but was so greatly modified that it became a distinct service style.

19. Food is plated in the kitchen by the chef. Plates are brought to the table by the server on a tray or cart. Guests are served from right to left, with food being served from the left, using the left hand and beverages from the right. These rules can be violated if they improve service. Bread is generally placed on the table in baskets for guests to serve themselves. In fine dining, guests may be served bread with tongs. It is not proper to remove any dishes or start a new course until everyone has finished, though this rule is violated in busy banquets. Items on the left are removed from the left with the left hand and items on the right, from the right with the right hand. In fine dining, crumbing is done after the entrée and before the dessert.

20. The American service style is the most versatile. It can be used for any meal period and in every sort of restaurant from casual to fine dining. It is fastest, requires the least space and allows for a very attractive presentation as the food is plated and garnished in the kitchen by the chef. American service requires less skill to perform than French or Russian.

Activity/Training

Take several menus and have students plan out how each would work using the five classic service styles. Of course modifications would be required. (It would not be necessary to serve Chinese food to serve food in the Chinese style, albeit without following each custom to the letter).

If there is a lab associated with the class, it would be instructive to plan to use an alternative to the usual, probably American style that students have been performing. Have students decide which method would work best or if a variety of methods could be employed, for example, beginning with appetizers served Russian style, followed by soup tureens brought to table and served English style, a salad tossed tableside French style, entrees served Chinese style on platters, and dessert flambéed French style, brought out on voitures or served simply plated, American style.

Chapter 8: Serving the Meal

Objectives

- Oversee proper setting of tables, proper meal service and clearing
- Describe receiving correct payment from customers based on accurate guest checks

Outline

I. Introduction

- The service profession has rules and procedures for accomplishing the required tasks of serving guests. This chapter reviews serving guests, tray and other equipment handling, as well as specifics for casual and formal dining.

- Isn't dining more casual these days? What about the Outback motto, "no rules, just right?" (It is true that customers are more comfortable with informality but that is referring to their own behavior- they want and expect good, smooth service whereby they are efficiently and competently served by a knowledgeable professional with a warm demeanor)

II. Steps in Serving

- There are five steps involved in serving food and beverages:

    - Greeting and Seating guests
    - Taking the Order
    - Serving
    - Clearing
    - Presenting the Check and Saying Goodbye

- Tasks of each step vary according to different meals and operations, desired pace of guests, and special nature of event meals.

- Ask students to go through the five steps above for: Pizza delivery, Mexican Quick-service drive-through, Family all-you-can eat buffet, French service style in a private club, and a private party at a nice, laid back seafood house with a view of the water. (Various as appropriate)

III. Greeting and Seating the Guests

- All employees should know how to welcome a guest as the first contact forms the first opportunity to make the experience positive for the guest

- Guests are typically greeted by the host/ess, owner or a greeter, who might be the server

- The elements of the greeting should include; pleasant attitude, warm smile, eye contact, and brief but welcoming phrase

- Guests with reservations should be honored with immediate seating but if not possible, they may have to wait a few minutes

- How do students feel about being handed a buzzer? (Various- adds convenience and lessens anxiety about being forgotten)

- Hosts/esses should be alert to accommodate guests seating preferences and pull out the chair for guests

- Special needs such as persons with disabilities and families with children should be addressed with helpfulness appropriate to the situation

- What kind of seating might a family with several young children prefer? An elderly couple with a wheelchair? A business dinner meeting? A romantic anniversary? (close to bathrooms, entertainment, with high chair or booster seats/, near the door for easy access, with room for the chair, not near cold air vent/, somewhere quiet and light without too many distractions/, somewhere private, not near the large family, a nice view or close to the dance floor/.)

A. The First Approach

  - Server's first contact is crucial as it sets the tone of what to expect. It should create a positive impression and establish the servers competence

  - The attitude should be friendly but dignified, do not indulge in unnecessary conversation or encourage familiarity

  - Servers should make sure their stations are ready to receive guests. Table settings should be appropriate to the establishment

  - Fine dining may or may not have salt and pepper on the table; condiments are supplied when guests order something requiring them. Casual dining may have tables preset with condiments.

  - Servers should be watchful for new guests being seated in their section. Guests must be promptly greeted, by name, if known, and acknowledged at least, no matter how busy it is

B. The Introduction

- The server may introduce themselves by name or simply as the server for the table when everyone is seated.

- Complimentary snacks should be immediately brought, if provided and water poured. Extra place settings should be removed after ascertaining that no latecomers will be joining the party

- Ask students if they want to know the name of their server and why? (Various)

C. Presenting Menus

- Menus are either handed out right away by the manager or host or following the premeal beverage order by the server

- Guests should be handed the menus either opened or not rather than having them (rudely) dropped on the table for guest retrieval. Menus should be handed to guests from the left, ladies first, if policy states

- Wine lists might be given by the server or separately by a sommelier

- Servers should describe all specials and their prices. Servers should have tasted all specials and be accurate and appealing in description and describe any additional items management wants to feature

- Do restaurants where students work/have experience with have servers taste specials every day? What kind of training is offered in regards to menu item knowledge? (Various)

- Servers should be familiar with all menu items, ingredients and methods of preparation

- Special guest questions about nutritional or dietary needs should be checked with the kitchen if the correct answer is not known

IV. General Rules and Procedures for Serving

- Proper service is a craft that, done correctly, flows so smoothly it appears simple to the untrained eye

- With sincere application in learning the basics and by acquiring dexterity in tray handling and dishware, the novice can become quite proficient and able to competently handle a good sized station.

- Servers cannot be considered professionals until they can handle all service equipment confidently

- Though the industry is constantly changing, servers should not assume that they can develop their own methods. Procedures and rules have been time tested and found to give the most satisfactory service

- The most basic personal requirement of a server is the ability to quickly get food and drink from order to the kitchen and back to the guests. This is a demanding task

- It is important for servers to observe the right method and not develop bad habits from the beginning

- All dining operations should establish a flow pattern that servers should follow when moving in the dining area and kitchen. Breaking the flow can cause accidents.

- Have students draw a flow chart of the lab associated with the class or other familiar dining room/kitchens. (various)

A. Serving Water and Ice

  - One of the first things done is to give guests a glass of water with ice. When pouring at the table, the pitcher should be 2-3" away from the glass rim.

  - Too close and the rim can chip, too far and a spill is risked. The glass should be poured 2/3-3/4 full

  - Buspersons and servers should be alert during the meal to see that glasses are kept filled. Some operations provide water pitchers or carafes on the tables for guests

  - Employees must know that spills require immediate attention. If in a busy area, an employee should stay and direct traffic, warning all

  - A "Caution, Wet Floor" sign must be posted while clean up is going on and until the floor is safe.

  - What kind of common injury is this practice designed to avoid? (Slip and fall)

B. Carrying Trays

  - Trays come in various shapes and sizes but the most common are the 27-30" oval used for large loads and the 15" cocktail try used primarily to serve beverages

  - Trays are customarily carried by the left hand raised slightly above the left shoulder

- For sanitary reasons, the tray should be at least 4" away from the head or neck avoiding contact with the servers hair

- When lifting it is important to plan a route, checking the floor and looking for hazards

- Employees should use their entire hand to grip the load, not just their fingers, keeping the load close to the body and centered

- Carrying procedures include;

    - Keep ears, shoulders and hips going in the same direction
    - Keep elbows against sides for additional balance and to ensure against bumping
    - Keep stomach muscles firm and lower back tucked in, load should be carried by the legs and hips, not back
    - To turn, move the whole body as a unit and avoid twisting

- It is not recommended to carry trays on the tips of the fingers unless needing to raise to get through a crowd- use the palm instead for firm support and better control

- Cocktail trays can be carried between the thumb and index finger with added help from the other fingers, particularly when carrying a tray of beverages for a party of six or more, tall stemware or large items such as coffeepots

- Items can slide on trays unless they have cork, plastic or rubber mat surfaces. A wet napkin can also reduce sliding.

- All trays should be kept clean for use

- Trays should not be placed on a table but instead a tray jack or stand should be used whether guests are at the table or not unless it is a large banquet or buffet once guests have gone

- While it may be impressive to carry glass or china without a tray, the tray must always be used for safety reasons

- Perfect balance is the secret of safe tray carrying. It can be obtained through even distribution of weight, placing the heaviest items at the center of the tray

- Jacks may be permanently set up or carried by the server with their free hand. If a tray is very heavy, the jack should be placed prior or a helper can come ahead and place the jack and help set down the tray. One or two dishes are removed at a time. When the tray is empty, it is removed.

- Loading trays requires attention to achieve balance and get a maximum safe load. Food covers are used to stack dishes and keep food hot. Eight dishes for dining room and 16 for banquets can be carried on one tray in this fashion.
- Beverages must be carefully loaded so the entire order can be brought safely in one trip.

- Why is one trip desired when it would be safer to do it in more? (Guests must wait for everyone to be served while food cools or ice melts, it also saves valuable time)

- Cold and hot items should not be placed on top of one another or even allowed to touch

C. Handling China and Glassware

- Fingers must not come in contact with the rim of dishes, stacked items must be held underneath and away from the body. A serviette or service towel may be wrapped around the stack to help hold it in place

- Tumblers and stemware should be held from bases and stems

- Chipped, soiled or cracked china and dishware must never be used

- Many dishes require underliners; juice, cocktail glasses, cereal dishes, soup bowls, dessert dishes and pots. Iced beverages should be set on an underliner or coaster.

- If items are hot, a serviette or napkin should be used to protect the fingers. The guest must be warned that the item is hot until they acknowledge the warning.

- Has anyone experienced being served a dirty glass or have a server put their fingers on the rim of a glass? (Of course)

D. Handling Flatware

- Never touch the part of flatware that will go into a guest's mouth

- Flatware should be wiped thoroughly before setting it on the table using a vinegar and water solution. Extreme spotting may require dishwashing procedures investigation

- Bent, tarnished or soiled flatware should never be put on a table

- Flatware should be properly set. When resetting, carry flatware on a tray or in caddies or on a plate with a napkin, never touching the part that goes in a guest's mouth

- The Steak N' Shake diner chain has a motto "In sight it must be right" to tout their open kitchen/grill. What desire does this fulfill for guests? (Sanitation- we know it is clean/sanitary as we can see what goes on. Sanitation practices must be kept whether guest is watching or not).

V. Taking the Order

- After premeal beverages have been served, guests should be asked is they have made selections. If guests have not, (maybe not even looked at menu) appetizer suggestions can be made with an offer to return

- Servers should be alert. Readiness to order is often indicated when a guest closes the menu and sets it down. Others may need assistance reading or understanding menu items

- Servers must be prepared to answer a host of questions including preparation times, ingredients, etc.

- Servers, or in formal settings, a Captain, takes the order. Often on a guest check which are numbered and signed for. Computerized POS systems have alleviated the need for the guest check process

- Servers should inquire if guests want separate checks or one and stand to the left of the guest ordering or in one place with a small group. The server should gather complete information and inform the guest if an item has a long preparation time

- Servers use systems to remember who got what. The pivot system is a common one whereby a common point of reference and the guest's proximity to that point clockwise around the table is used to identify orders.

- Discuss the system used in lab setting if applicable or student experience with systems.

- Computerization allows coding in of guest locations, time order is placed, ready and time picked up so management can check on the flow and find source of delays. Computer system will also print sales tickets.

- Handheld systems are growing in popularity but still require a system such as the pivot system for taking orders

- In guest check systems servers write out orders legibly, clearly and using standard abbreviations. It is helpful to underline extreme doneness and note requests

- Always repeat the order after the guest gives it to prevent errors and misunderstandings, getting complete information. This may require prompting. Dessert orders are usually taken after everyone finishes with the entrée.

A. Suggestive Selling

- Servers have an important roles as salespersons
- Suggestive selling helps guests come to a pleasing decision and is an essential part of excellent service

- Servers must ask questions and watch for clues to help guide guests who rely on servers to help them make the right decision

- Appropriate attitude, dress, and confidence will encourage guests to take your suggestions.

- Don't be overly aggressive, causing resentment and mistrust but don't be afraid to make helpful suggestions. The bigger the check, the bigger the tip.

- Open ended questions allow further discussion or suggestion of options.

- Knowing the menu and reading the guest will help as well as the suggestions:

    - Beverages and appetizers to start the meal
    - Premium liquors when house brand is ordered
    - Fresh fruit if hesitation on dessert
    - Side orders with entrees
    - Make definite suggestions versus "anything else?"
    - Know menu and suggest low calorie when appropriate
    - Suggest an appetizer if order has long preparation time
    - Complete dinner if ordered a la carte
    - Use appetizing words to describe
    - When items are available for takeout, suggest take home order
    - Desserts, sharing desserts and after-dinner drinks

- Give class some scenarios and have them role play suggestive selling. Have students "guests" play the roles of: indecisive big spender, very hungry but uncertain, on a tight budget, on a low calorie diet, and knows exactly what they want and resents suggestions. Coach them on the various situations

B. Placing and Picking Up Orders

- Computerized ordering which often has a preset keyboard or touch screen does not require servers to go to the kitchen and may prompt servers for order specifics such as the doneness of meat

- With manual systems server still place their orders in the kitchen. In fine dining the order may be checked by a checker who places the order, in formal dining, the aboyeur or expediter may take the order form the server and place it at the appropriate stations.

- Servers still call out their orders in some settings but this can be confusing and noisy. Quieter methods are used, such as the spindle method, where written orders are placed on a spindle. Wheels are also used to keep the orders in sequence

- Various methods are used to notify servers when their orders are ready. Pagers, lights or numbered lights can be used.

- Before taking the order out, the server must check to see everything is correct (checkers sometimes do this) Servers should never take out something that does not seem right

- A course should not be brought out until the previous course has been finished and cleared except in some faster service operations

- Pickup may require getting food from various stations. Good organization simplifies the service task at the table and ensures correct temperature, pleasing appearance and superior quality. Expediters sometimes act in this capacity.

- Discuss the consequences of taking out the wrong order (Two tables are disserved; kitchen may need to prepare some item again needlessly, food cost, waste, guest dissatisfaction etc.)

VI. Serving the Guests

- Serving the order is the total of all efforts of everyone involved in the operation- none of this should be lost in the dining room

- Most foods, unless they are shared appetizers, are placed directly in front of the guest with bread or side salads placed to the right and left respectively. The main entrée item should be directly in front of the guest (6:00 position)

- Beverages are placed to the upper right with cups and saucers to the right with the handle to the right

- There is much variation today but the important thing is to make things convenient for the guest. Managers setting standards should provide service guests want and servers are able to meet

- Servers should use their left hands when on guests left and right on right, never reaching in front of or across guests

- Decorations, salt, pepper and other condiments are usually in the center of the table except for booths or table on a wall where they are against the wall. Bread trays and baskets are placed in the center

- Ask students if they have ever been struck with table arrangements that were particularly pleasing or not so. Ask them to describe these settings and analyze. (Various)

A. Breakfast Service

    - Service must usually be fast

    - Breakfast food particularly loses quality if not served immediately. Hot beverages should be served very hot

    - Many operations offer buffet breakfasts allowing guests to move quickly and get exactly what they want. Most offer a la carte in addition for lighter eaters.

B. Lunch Service

    - Lunch may consist of only one dish, served directly in front of the diner. Side dishes, if ordered are: vegetables to the right, salads to the left. Beverages are to the right with handles of cups turned to the right at 3:00.

    - Tables are crumbed at formal luncheons between courses

    - What kind of formal lunches might take place in today's world? (Meetings, occasions such as showers or parties, awards etc.)

C. Formal Dinner Service

    - The first course in a full-course meal is placed directly in front of the guest with specific utensils on the right

    - Fingerbowls offered after course should contain warm water, lemon and a fresh napkin

    - Soup is a typical second course. The bowl is placed on a serving plate with the spoon on the right. Crackers are offered, with crumbing done as needed between courses.

    - Entrees are placed in front of guest with vegetable dish above and to the right and salad to the left. If guests serve themselves from a platter, a warm dinner plate is placed in front of the guest

    - Formal dinners may serve salad after the entrée placed directly in front of the guest on a plate or in bowl

- Fresh fruit sorbet served before the entrée is usually served in a tulip champagne glass with doily or underliner.

- Dessert utensils are set above the plate or brought with dessert. Pointed desserts such as pie should have the point facing the guest

- Were any points of service unfamiliar to students? Did any seem stiff, or unnatural? Discuss.

D. Booth Service

- Often difficult, service to guests at the far end should come first. Students must lean and reach- using left hip and hand for those on the right and right hip and hand for those on the left. Items may need to be cleared as early as possible as booths may have less space than regular tables.

- How about just handing off dishes to those seated nearby? (Customarily, guests should be treated as though they are "armless" and not on the payroll. Naturally, one should not get into a "tug of war" with a guest who "insists" on helping out; however, guests should never be enlisted.)

VII. Clearing Tables

- In some operations, clearing is not done until all guests have finished the course, in others, it is acceptable to clear as guests finish. Guests who ask, or appear to be uncomfortable with dirty dishes should have them removed, regardless. When in doubt, ask.

- Clearing is done from the right, using the right hand

- Unfinished beverages should be left unless guests want them removed. Offer after dinner drinks; leave hot tea and coffee for the dessert course. Water glasses remain until the guest leaves, keeping filled

- Table should be crumbed, whether or not it has a tablecloth using crumber, dinner knife or rolled napkin (not onto floor). The goal is to tidy table for dessert etc.

- Why are dirty dishes left (sometimes) until everyone is finished? (It is considered rude to make those still eating feel that they have been too slow and that they should rush or not finish)

- Is it ever acceptable to instruct guests on manners? (If a guest asks what something is for, etc. they should be answered politely but a guest doing something "rude" such as asking to have dishes removed before everyone is finished should never be told that they are committing a blunder. It is ruder to inform someone that they are ill mannered than it is to be ill mannered as manners are about making people feel comfortable)

VIII. Presenting the Check and Saying Goodbye

- Guests want the check when they are ready to leave. Server and host/ess should be watching for signals of readiness to leave.

- Difficulty getting the check can cause ill feelings, leading to a poor tip
- Present the check after the last clearing and asking if anything more is desired.

- The accurately totaled check can be presented in two ways:

    1. Brought to the table and placed facedown with thanks, an invitation to return and instructions to pay the cashier or:
    2. Brought to the table and placed facedown with thanks, an invitation to return and departure by the server so that the guest can scrutinize the bill privately. Change, if desired, or credit card processing is done by the server

- It is common to bring a receipt. Manipulating the amount is considered to be fraud

- Why would the amount be manipulated? (Guest wishes to claim more expenditure than actually occurred to employer etc. Often a blank receipt is brought so that the decision to be ethical or not is one that the server is not involved in.)

- Personal checks generally require approval procedures including an identification check

- Credit cards are much more common form of payment requiring the guest to sign the authorization slip, matching name and signature to that on the card

- It is unprofessional for the server to show that a tip is expected or to express disappointment over a tip amount

- In Europe, tips (gratuities) are often included automatically, as may be done in the U.S. for large groups.

- While a poor tip may indeed represent poor service, it may also represent a miserly guest or one not familiar with our customs. Is it fair to the server to be so dependent on guests for what may be the bulk of their incomes? (various)

- Guest checks should be used for all sales transactions by all who make sales, ordered by management, secured and issued in recorded lots subject to audits

- Point of sale systems reduce or eliminate the need to have numbered checks as they produce checks using data entered by servers. They may also process credit card payments

- Departing guests should always be thanked and invited to return, leaving with a good feeling

- Why are the greeting and departing stages considered to be so important? Isn't *during* the issue? (First and last impression and, of course)

IX. Closing

- When serving has ended there are closing duties;

    - Soiled linen should be counted, bagged and properly stored
    - Unused linen should be folded and stored
    - Unused clean flatware etc. returned to storage
    - Condiments wiped clean and properly stored
    - Salt, pepper, sugar and other items placed on trays to be replenished
    - Butter should be covered and stored under refrigeration
    - Work station should be placed in good order
    - Tables needing fresh linen and resetting should be done
    - Sidework should be completed as assigned by management

- Closing after a serving period varies according to whether another meal period is coming, the day is ending and what the next meal will be

- Review the closing procedures and Sidework for lab associated with the class, or have class brainstorm on; What Sidework could be done by the first, middle and last to be released from a shift? What Sidework can be accomplished while guests are still present and which could cause guests to be disserved? (Various answers)

X. Formal Dining

- Formal dining is characterized by a number of factors including:

    - Stricter, more formal dress code
    - Elaborate food and service
    - Elaborate menus
    - Elegant decor
    - More servers used
    - Reserved, quiet, peaceful atmosphere
    - Prices reflecting the cost of these extras

- Basic service mechanics are applicable but execution is more elaborate

- French service may be used with the gueridon, Russian or American

- Wine is featured much more as a meal accompaniment; a sommelier is usually on staff providing more formal, elaborate wine service

- All dining room servers must be well trained and work might be specific to positions (team service concept)

- How often and where is this sort of service and atmosphere seen? Do guests want fine food in a more relaxed, casual setting? Name the "best" restaurants in your geographic area and describe the atmosphere and service. (Various- surveys do seem to indicate that guests want a quality experience in a more relaxed setting).

XI. The Busperson's Role

- The busperson is often the employee after the server to make contact with the guests. Many are young but some keep the position as a career. This professional position can lead to job satisfaction

- Should class members snort at this proposition, I am reminded of a fine dining restaurant I worked at with teams consisting of; Captain, Front and Back waiters and Busperson. We had a head busperson on staff (Carlos) who was very competent. When offered the chance to move up to Back waiter with a Front position soon to follow and the possibility of later promotion to Captain, he turned it down, citing job satisfaction and income of his current position (he got a percentage of tips from all buspersons due to being in charge of them) I can't imagine that this is the only of such examples.

- Busperson is responsible for providing water; butter and perhaps, bread after guests are seated and have menus.

    - Bread baskets should be lined with a napkin

    - Butter should be cold but not on ice

    - Water is generally poured at the table but water in glasses can be brought on the cocktail tray

    - Other duties may include the following:

        - Carry food from the kitchen
        - See that ice and water are at hand
        - Assist in removing soiled items in appropriate manner
        - Provide supplementary service items requested by guests
        - Assist in beverage service
        - Refill coffee and tea orders
        - Ensure condiments and service supplies are available
        - Help maintain linen, equipment and service ware
        - Be available at all times for guest's special requests
        - Reset tables after guests depart

- In order to do the job effectively, buspersons should follow these guidelines:

    - Keep equipment organized
    - Avoid over stacking which causes breakage/accidents
    - Keep safety in mind always
    - Walk, don't run, perform systematically
    - Try not to travel empty handed
    - Do not engage in long conversation with guests unless encouraged
    - Good communication skills with servers and kitchen workers help job to flow more smoothly
    - At the end of the shift, leave station immaculate

- Buspersons should remember to anticipate guests' needs. Do not make patrons have to ask for anything

- The buspersons approach should be immediate and courteous

- Class could write job description for a busperson based on the class lab, a fine dining room with formal service, a California menu restaurant, open 24 hours, an all-you-can-eat buffet etc. (various)

Chapter Review Questions Answer Key

1. Heavy items are placed in the center of the tray and lighter items on the periphery so that the weight is evenly balanced.
2. When loading hot and cold items, it is important that they are not stacked on top of each other and in fact, should not even touch.
3. Tray surfaces are often covered with cork to prevent dishes from sliding and spilling or falling and breaking. If the tray is not equipped with cork, a wet napkin can assist in avoiding slips and spills.
4. The host or hostess normally first greets guests. In other situations the owner may be stationed at the door, greeting guests or a person occupying the position of greeter.
5. Depending on operation policy, the server should either come to the table, once diners have been seated and announce, in a friendly tone, "Hello, my name is:_______, and I'll be your server this evening" or simply, Hello, I'll be your server this evening, welcome to:______".
6. Servers help guests through suggestive selling by assisting them in coming to a decision about what they want. By asking open ended questions and eliciting information, the server can guide the guest toward an enhanced experience. Suggestive selling should never be overly aggressive or the experience is not enhanced. Servers help themselves when guests are satisfied and leave a nice gratuity on a bill with a higher total than if no suggestive selling had occurred.
7. The abbreviation for steak, medium rare is: St. MR

8. Today, much order taking is computerized. Servers send orders through the computer in the dining room to cooks in specific kitchen stations. It is possible for servers to code in cover locations and the time the order is placed, ready and picked up so management can check on the flow of service. The computer system also will print sales tickets. A POS system often has a preset keyboard or touch screen with almost all items on a menu listed and prompts for further information needed such as the doneness of steak.
9. China, glassware and flatware should always be separated. Soiled items should be stacked according to shape and size to allow more room and give proper support to stacks. It allows more items to be stacked on a tray and saves trips and also helps the dishwasher.
10. If the guest is to pay the cashier, the check should be presented after the last clearing has occurred after first courteously asking if the party desires anything else. The accurately totaled check may be brought to the table and placed face down, with thanks and instructions to pay the cashier. If the server collects payment, the check is brought to the person paying the bill and is presented face down, by the departing server. The guest leaves payment plus a tip and leaves. It is common to bring a receipt to the guest.

Chapter Cases Answer Key

Putting on a Banquet

This is a case of suggestive selling. The manager must first address the guest's concerns. The guest is worried about congestion; this may be addressed by showing him a layout of the room and giving him references from others who have been satisfied with this form of service. He thinks many members may not like the arrangement. He might be further queried as to why he believes this to be the case. Are they elderly and may not wish to stand? A few chairs or sofas could be provided, if this was the case. Does he feel that there would not be enough service? Demonstrate how guests are able to serve themselves exactly what they want, being as indulgent (or not) as they would like from a lovely display. Additionally, the manager can explain that the guests will have been seated with the same 7 people (or three couples) and that this arrangement allows guests to get up and circulate, socializing with more people than were simply at their table. Each guest is free to go where they please, eat what they please and speak to whom they please. Those who are ready to call it a night can slip away easily. The president can be questioned to see if there are any further doubts to be addressed but must be allowed to choose what he wants for his party.

Handling a Personnel Problem

Teamwork is important and the team has been compromised. First of all, quarrels that are nearly fights should each be addressed as they occur. Participants should be warned about unprofessional behavior and allowing their personal lives to interfere with their job duties and smooth functioning of the team. If necessary, verbal or written warnings may need to be issued and placed in participating employees' personnel files.

It might be wise to meet one on one with the waiter who is the cause of the problems. He too should be instructed on professionalism and the problem with letting personal issues affect the teams' ability to perform. He should be commended for the fine work he does and contributions made.

There are operations that have a policy against employees dating each other. It seems to overstep but the intent is to avoid situations like these. It might be good to have a general meeting where employees are reminded to keep their personal lives personal. Behavior or actions which affect the job or team work on the job is subject to disciplinary action. If there is no policy prohibiting personal plans, romances and friendships that interfere with job performance, it may be time to put one in place. While no employer can dictate what an employee legally does during time not on the job, they can dictate behavior on the job.

It is also possible that strategic scheduling could help cool things down. To reward those whose behavior has been professional, they might be assigned preferred shifts, stations etc. If problems continue, so should warnings with those who are unable to focus at work let go, if need be.

Chapter 8 Test

True/False

T F 1. Service is not really a profession; it is a job on the ladder to restaurant management.

T F 2. Extra place settings should remain on the table to be used by the next party if there are any seats that are not occupied.

T F 3. If a server does not know the answer to a guest question, they should make their best attempt to answer the question in order not to appear to be incompetent.

T F 4. All dining operations should establish a flow pattern that servers should follow when moving in and out of the dining area and kitchen.

T F 5. The professional server has learned how to carry tall stacks of dishes against the body and stemmed glassware between the fingers.

Multiple Choice

6. Serving food and beverage involves a sequence of five steps: (1) Greeting and seating guests, (2) taking the order, (3) Serving, (4) Clearing and (5):
   a. Crumbing
   b. Dessert and Coffee Service
   c. Presenting the Check and Saying Goodbye
   d. Closing duties including Sidework

7. The first employee that a customer comes in contact with:
   a. represents the first opportunity to provide a positive experience for the guest
   b. is the person taking the reservation
   c. is the valet parking attendant, or, if not, the host/ess
   d. is the server, who should introduce themselves and take the beverage order

8. Menus should be handed out by:
   a. placing them in the center of the table, to be retrieved by guests when they are ready for them.
   b. the first person to notice that the table is without them
   c. handing the menus to the guests from the left at a time established by operation policies and procedures (after beverages, or as the party is seated)
   d. giving them to the host of the party to pass around, thus establishing who is to receive the check

9. Recommended service procedures are:
   a. So hopelessly out of date as to be irrelevant
   b. Tested over time and found to give the most satisfactory service
   c. Subject to the modification of each server
   d. To be rigidly adhered to at all times

10. Trays should be carried:
    a. High overhead on the tips of the fingers
    b. By lifting them from the waist, employing the muscles of the lower back
    c. By the left hand, carried slightly above the left shoulder
    d. Carefully and placed smoothly upon the table to be served

11. The server ordering system whereby the guest with their back to a specified point in the dining room being numbered cover 1 or A followed clockwise is often called the:
    a. Point of Sale system
    b. Pivot system
    c. Check back system
    d. Guest-check system

12. Abbreviations used for ordering should be those:
    a. Most easily understood by the guest
    b. Most easily remembered by the server
    c. That have been standardized by the operation
    d. That fit most easily onto the order pad

13. The meal that is usually fastest with the most rapid deterioration of food quality when cooled is:
    a. Drive-through
    b. Breakfast
    c. Buffet
    d. Formal dinner service

14. Food should be served from the ____ using the ____ hand, beverages from the ____ using the ____ hand?
    a. right-right, left-left
    b. left-right, left-right
    c. right-left, right-left
    d. left-left, right-right

15. Clearing is done:
    a. After all guests are finished or when guests indicate that they wish to have dirty plates removed
    b. By the host/ess, when not seating and greeting guests
    c. When the party has paid and left the table
    d. As the next course is being served

Short Answer/Essay

16. If the server is too busy to attend to a table that has been sat immediately, what can the server do to acknowledge the guests?

17. What happens when a server develops the wrong habits from the beginning? What are the "right' habits when it comes to rules and procedures for serving?
18. What is the difference between suggestive selling and aggressively annoying the guests?
19. Describe the characteristics of formal dining.
20. Comment on the statement: "Good service means responding quickly to all guest requests".

Chapter 8 Test Answer Key

True/False

1. False, service has rules and procedures for accomplishing required tasks as does any profession. A great number of people have found lucrative, satisfactory careers in the service industry as servers.
2. False. Extra place settings should be removed in order to give guests more room and for sanitary purposes.
3. False. The server must seek out the correct answer, including asking the chef.
4. True.
5. False. These habits are risky and unsanitary. Plates should be carried with both hands underneath, away from the body and glasses should be carried on a tray.

Multiple Choice

6. c
7. a
8. c
9. b
10. c
11. b
12. c
13. b
14. d
15. a

Short Answer/Essay

16. The server can make eye contact and briefly welcome the guests and let them know they will be attended to as soon as possible. It is important to present a calm demeanor, rather than a harried, frantic one or the guest will not feel they can trust in the competence of the server.
17. Servers who develop wrong habits become accustomed to them and provide less than adequate service. The right habits are the policies and procedures developed over time and found to give satisfactory service. The management of the operation should train servers on how to best serve the facilities guests.
18. Successful suggestive selling helps the guests reach decisions that will enhance their experience. Aggressively annoying guests is failing to read them and relentlessly making suggestions for the sole sake of raising the total of the bill. Guests resent this and rightfully feel disserved.
19. Formal dining has a stricter, more formal dress code. The food, service and menus are quite elaborate. The décor is elegant with a quiet, reserved, peaceful atmosphere. There are more servers and the prices reflect the higher cost of these features.
20. While it is true that guest requests should be responded to quickly, the best service is provided when guests' needs are anticipated and they do not need to request anything.

Excellent servers and buspersons are ever alert to guest needs and are trained in what utensils, condiments and service is needed for each menu item.

Activity/Training

Class can be assigned as individuals or in groups to develop a service critique sheet for a variety of kinds of dining operations, based on principals in this book and their own preferences. They can then perform a "mystery shop" using their critique sheet and score the establishment. Having them rate dimensions on a scale of 1-5 with "5" being the most favorable helps to come up with a numerical score. A space for explanatory comments after each dimension is helpful as well as categorizing rating to pinpoint strengths and weaknesses. The student(s) can then develop a training program to address deficiencies.

Chapter 9: Bar and Beverage Service

Objectives

- Know enough to serve and recommend beverage alcohol and nonalcoholic beverages to guests based on informed knowledge of beverages
- Correctly follow legal and ethical procedures for serving alcoholic beverages

Outline

I. Introduction

- People have concocted alcohol for 1,000s of years from fermented milk, berries and fruits and mixing water with essences of herbs, seeds and spices

- References throughout history have in included cave pictographs, a stone tablet near where Noah's ark was said to have landed and grape fermentation 8,000-10,000 years ago

- Tequila was made from agave cactus and honey mixed with fermented liquid was used to make mead. Ciders were poured into open containers to ferment

- Coffee and tea were consumed by early civilizations in southern Asia where coffee was first discovered and thought to have medicinal properties

- Overuse was found to lead to indigestion, depression, irritability and anxiety. This and other reasons led Pope Cement VIII to declare coffee an infidel drink and thus forbidden it was later tasted by another pope and "baptized" as acceptable

- Today it is the most widely consumed beverage

- Is coffee running its course in popularity? If so, what would replace it? (Probably not, specialty caffeinated beverages; tea, chocolate etc.)

- The Chinese were the first to drink tea made from the leaves and buds of a semitropical bush related to the camellia

- Coca Cola was first made in the 1880s but carbonated beverages were not mass produced until the 1920s and 30s and are today internationally popular with the per capita amount in the US increasing

- Carbonated beverages are often cited as being empty calories. Is a new generation of more nutritious beverages taking the place of the old standbys? ("Natural" soda, carbonated juice and vitamin enriched beverages, teas etc. are making inroads but have not replaced sodas)

- Specialty spirit drinks and wine are very popular
- Premium and specialty beer sales grew in popularity throughout the 1990s and remain strong sellers
- Beer is by far the most popular alcoholic beverage and in 2003 made up more than 50% of all alcohol sales
- Why is beer so popular? (wide spread appeal, affordable, popularity spans age groups, associated with sporting event/"stadium" consumption, college aged students etc.)

II. Some Important Facts about Beverage Alcohol

- Alcohol forms when yeast breaks down carbohydrates, freeing carbon dioxide, adding ethyl alcohol, a depressant and intoxicant.
- Alcohol is measured as blood alcohol content or BAC in people's systems. Most human livers can only break down one ounce of alcohol per hour.
    - BAC of .05 can reduce inhibitions, reasoning and slow down reactions
    - BAC .08 is legal limit indicating intoxication and often includes slurred speech and blurred vision
    - BAC .15 loss of muscle control and alcohol acts as poison causing vomiting and nausea
    - .3-.4 person passes out
    - .4-.5 deep part of brain is affected, loss of essential body functions and death
- Moderate consumption of alcohol has been shown to have beneficial effects; linked with lower risk of heart disease and lower blood pressure, is relaxing, and stimulates the appetite
- Ask class if they prefer to work in restaurants that sell alcohol or those who do not and why. (Various- challenge to deal with clients drinking, open later, more money as associated with fine dining, etc.)

A. Alcohol Content and Proof

- Alcohol content must be stated on the label by federal law
- Proof is weight content- proof is double the % (ie 60 proof = 30%). Proof is used for spirits

- Wine and beer indicate alcohol content by volume but this overstates the amount by weight as alcohol is lighter than water (80:100) therefore, beer of 6% and wine of 14% are 4.8 and 11.2% alcohol

- 1 ounce of pure alcohol = 2- 12 ounce beers, 1- 8 ounce glass of wine, 2 ½ ounces of 80 proof spirit and 2 ounces of 100 proof spirit

- Does the (relatively) low alcohol content of beer contribute to its popularity? (Various- more associated with thirst quenching than other forms of alcohol, typically drunk in larger quantities (pitcher) than wine or spirit alcohol. Beer can be paired with food, much like wine.

B. Aging

- Spirits and wine are improved by age. Beer does not and should be consumed as soon as possible after production.

- Wines age in the cask and after bottling

- Spirits age only in the cask- there is little change after bottling

- Some names are indicative of the length of time a product is aged

III. Knowing Spirits

- Servers need to know about alcoholic beverages so they can inform guests and provide proper service

- Spirits are produced by the fermentation of grains, fruit, plants and other products

    - After fermentation the brew is distilled at or above 160 F
    - Alcohol vaporizes leaving the water in the brew, when vapor is condensed, it is mostly alcohol
    - Alcohol carries over flavor of the products from which they are made or added after distillation

- Spirits in the US cannot be less than 80 proof, many are higher. Cordials can be less

- Bonded spirits must be produced from a single distillation at 160 proof or less, be bottled at 100 proof and be unblended (it is a straight liquor)

A. Whiskey

A. Straight whiskeys are from one distillation, a Blended whiskey is a straight whiskey blended with other batches of whiskey or grain neutral spirits

B. Bourbons must be made from at least 51% corn as fermentable material and normally be aged four years in charred oak barrels at 125 proof

C. Rye Bourbon is distilled from a mash containing not less than 51% rye as the fermentable material. It is like bourbon but heavier in flavor. It must be aged at least two years in charred oak barrels

D. Corn Whiskey must be made from a mash containing at least 80% corn

E. Canadian whiskeys are blended and are lighter than American whiskey

F. Scotch whiskey is made from malt or grain with the smoky flavor coming from drying sprouted barley over peat fires. American whiskey must be 80-86 proof and aged at least four years

G. Irish whiskey is made from barley but with no peat in the process, thus without the smoky flavor. It is usually 86 proof and aged at least seven years; it has a heavier body than Scotch

H. Would alcoholic spirits have standard recipes? (Of course- product consistency, etc.)

B. Liqueurs (Cordials)

- They are distilled spirits that have been treated with special flavoring ingredients
- Many are quite sweet, a distinguishing characteristic
- Some have been made for centuries. Cordials have been classified by their flavoring ingredient
- Why would recipes be kept secret, by monks, for centuries? (Profit, pride, all speculation)

C. Gin

- Gin would be grain neutral except that usually in the last distillation, botanical flavoring substances –Holland so much so that they are not good for mixing.
- Few gins are aged, if they are, they take on a golden color and are called golden gin

D. Vodka

- Vodka is neutral spirits treated after distillation with charcoal and other materials as to be without distinctive character, aroma, or color, typically between 80-100 proof

- Vodkas have enjoyed an increase in popularity due to the production of flavored vodkas; citrus, berries, and spices

- Does flavoring a drink prized for its neutrality change the very nature of the product? (It does seem to be a contradiction)

E. Tequila

- Distilled spirit made from the core of a mature blue agave plant where starches turn to sugar through a roasting process, juices are extracted and fermented in tanks. It is 70-110 proof. The worm is mostly a marketing ploy.

- Why would a worm make a product seem more desirable? (Myth, thought to produce hallucinatory properties in the beverage)

F. Rum

- Sugarcane or molasses are the sources of rum

    - Light-bodied rums (silver or dry) are smooth and have a light body and slightly sweet flavor and are usually aged one year
    - Heavy bodied rums are darker in color and richer in flavor with tan or brownish labels
    - Why are rums associated with the Caribbean and Gin with Europe? (ingredients found geographically; sugarcane and juniper berries respectively)

G. Brandy

- Any spirit distilled from a mash of fruit or fruit derivatives can qualify as brandy.

- Only a distillate made from grapes can be called brandy- other fruits must be labeled by the name of the item used such as; Apricot Brandy

- Cognac must come from the Cognac region in France; Armagnac comes only from that French region. Calvados is an apple brandy from Normandy.

- Most brandies are aged, some for very long periods. Most are 80-100 proof.

- Labels have indicative symbols; VS or 3 stars means brandies are less than 4½ years old, Vieux or Grand Reserve means most are 20-40 years old and the youngest is more than 5½ years old.

H. Grain-Neutral Spirits

- Many alcoholic beverages are made from grain-neutral spirits, a plain unflavored liquid made of pure alcohol plus water. A grain-neutral spirit is the alcohol used to make alcohols that are *blended*

IV. Knowing Wine

- Wine is made by fermenting fruit; yeast changes the sugars to alcohol and carbon dioxide, while extracting fruit flavor.
- Wines are not distilled
- Aging can improve the flavor of wines- all wines are aged a year in the cask before they are bottled, after that, aging can continue
- White wines age well, usually up to about six years, but some age longer
- Red wines can age for many decades with dramatic change in flavor, color and body
- Wine alcohol content can be from 6-20%, if above 14% the wine usually has a spirit added to it

A. Types of Wine

- Wine can be categorized as three types:

1. Table wines- they are non-bubbling or still. The C02 gas byproduct of fermentation is allowed to escape. Alcohol content is 8-14%. They are sold by the glass, bottle and carafe

2. Sparkling wines- bubble because yeasts and sugar are added to still wine, causing a second fermentation in the bottle. They can be white, red or rose (blush) and contain 8-14% alcohol plus C02. The best known is Champagne.

3. Fortified wines- are those with added alcohol, usually brandy. All wines with more than 14% alcohol have been fortified. The two varieties are:

   - Aperitif wines- flavored with herbs and spices and served before a meal (Vermouth)
   - Dessert wines- are typically sweet, heavy, and commonly served after dinner (Port, sherry)

B. Grape Varieties

- Red wine is made from grapes with red, black, or purple skin fermented with their skins on giving their color and tannin (strong slightly bitter flavor that makes the mouth pucker)

- Red wines tend to be heartier, heavier and more flavorful than whites and roses and are almost always dry, or lacking in sweetness

- White wine can be made form red, green or white grapes. The juice is extracted and separated from the skins then fermented.

- White wines are lighter in color, flavor and body then reds and generally have a shorter life span. They can be very sweet or very dry and range from pale yellow to gold, some having a greenish tinge

- Blushes or roses are pink and range from very pale to nearly red. They have a fruity flavor and are light and fresh; many have a slightly sweet taste.

- Blushes can be produced by fermenting grape skins for a shorter duration or blending red and white grapes. Some are called white, such as white Zinfandel and white Pinot Noir from California.

C. How Wines are Named

- In the United States, wines are named in three ways:

    1. By the predominant variety of grape used (varietal)

        - The name of the grape is the name of the wine (Cabernet Sauvignon, Chardonnay, and Zinfandel)
        - These terms were introduced to differentiate American wines from European wines.
        - Legally, a varietal must include at least 75% of the dominant grape

    2. By the broad, general type (generic)

        - Though they are American, names are borrowed from European wines that come from well known districts. They have a resemblance to the original.
        - By law, all American generics must include a point of origin on the label
        - The best are pleasant, uncomplicated and affordable and often serve as house wines. Wineries use terms like red or white table wine for these blends

3. By brand name

    - Can be anything from an inexpensive blend to a very fine wine
    - Another name for brand name is proprietary name. In France it is called a monopole, belonging exclusively to a vineyard or shipper who produces and/or bottles the wine and takes responsibility for the quality. Reputation is the guide.

- Imported wines may also be named by their point of origin

- Various bottles of wine (empty or not) could be brought in to illustrate wine labels with students identifying which type of wine each is.

D. Tasting Wine

- When tasting wine, a stemmed glass should be used, held by the foot or stem, never bowl with the same (no more than half full) amounts of each wine should be poured for comparative purposes

- There are generally considered to be five steps involved in wine tasting:

    1. Hold wine up to the light to show color

    2. Next, swirl and smell. Do not swirl sparkling wines as it releases their distinctive bubbles

    3. Smelling should be done with the nose inside the glass taking deep breaths. Aroma is the scent of young wine, aged wine becomes the bouquet.

    4. Actual tasting determines flavor and other sensations. The wine can be swallowed or spit out after tasting

    5. The impression left in the mouth is called the finish. This is the evaluation period of a wine tasting

- If feasible, have a purveyor conduct a tasting for the class, attend one at a wine store or restaurant or conduct your own.

E. Pairing Wine with Food

- First wines and foods in each operation must be tasted so they can be described and recommendations made. Individual tastes vary so general rules are only suggestions.

- Appetizers- dry to medium dry, light bodied acidic white wines stimulate the appetite. Sparkling also work with hors d'oervres

- Fish and Seafood- dry white wine allows subtler flavors to surface

- Poultry and Pork- depending on sauce, a white, blush or red may all be appropriate

- Veal- complimented by lighter red, well-aged red or dry white wines

- Beef and Lamb- high fat requires wines such as complex reds with tannin to cut through full flavor. Seasoned or marinated meats need a younger red.

- Ham-sweet saltiness needs careful matching. If glazed, fruity rose or blush

- Ask students who drink wine what type they like to order and if they base their selection on food, server suggestion, price etc. (Various)

V. Knowing Beer

- Beer is the largest-selling beverage alcohol

- Beer is made from grains, yeast, water and hops. Rice and wheat are sometimes used

- Malt made from sprouted barley is the grain most commonly used

- Sprouting turns the barley's starch into sugar called maltose. The sweetness of malt is largely lost as sugar changes to carbon dioxide and ethyl alcohol with the exception of beers such as porter and stout

- Special yeasts are needed to produce different kinds of beer.

- Beer is more than 90% water and brewers claim excellence stems from water used

- Hops give a pleasant better flavor

- The most popular style in the US is pilsner made by the lager method which requires a long period of cold fermentation, proceeding slowly from the bottom. Lagers are about 4% alcohol by volume

- Top fermented beers require only about three days to ferment (ales, porter and stout)

- Porter, stout and bock beers have alcoholic contents (6%) and are heavier, richer, darker and sweeter than lagers

- Malt liquor has the highest alcoholic content of beers
- Light beers are popular due to about 100 calories for 12 ounces versus 140-175 for other beers
- Brewers are constantly striving to create new types of beer
- What kinds of beers are offered/popular at various establishments in the class community?

VI. Knowing Nonalcoholic Beverages

A. Coffee

- Coffee has long played a major role in food service
- The coffee tree (Coffea Arabica) is a tropical evergreen shrub native to Afria.
- The beans produced by the coffee tree include Robusta and Arabica. The word coffee is thought to have originated from the Ethiopian region of Kaffa
- Jean de la Roque reports that the Turks made the beverage popular in the Middle Ages. De la Roque introduced coffee to Europe bringing beans and brewing the first cup of coffee in Marseilles, France, 1644
- The first coffee houses on records are found in old Constantinople (now Istanbul)
- Roasted coffees are sold as one type or blended. Light roasting produces a mild flavor while long roasting creates a dark, stronger coffee. After roasting, the beans are ground.
- The fineness of the grind has much to do with the extraction time- the length of time the water and the grinds are in contact. Fine grind time is two minutes after which grinds should be removed to prevent the bitter remnants of coffee from harming the flavor
- Coffee flavors are carried in bean oils which are fragile and oxidize easily. Ground coffee must be kept vacuum packed or frozen. Whole bean coffee keeps its flavor for two weeks at room temperature
- Some food service establishments buy fresh roast whole bean coffee and grind each brew independently to produce high quality coffee
- In the past two decades, decaffeinated coffee has become increasingly popular. In many operations, it is requested 2:1 over caffeinated. Decaf contains only 3%

through an extraction process done with chemicals or the superior European water extraction method.

- Coffee is often served last and thus leaves the last impression with guests. Coffee follows the entrée unless requested earlier, usually before or with dessert.

- Might that 2:1 ratio be due to the fact that dinner guests are more likely to be concerned with caffeine than breakfast guests? Would the ratio of 2:1 hold for breakfast or California menu restaurants? (doubtful- statistics said "in many operations")

- Before coffee service, table should be cleared and crumbed. In a coffee shop or family style restaurant, it is served in 6 ounce cups.

- Coffee should be served hot (above 160 F) to give the best flavor

- Scrupulously clean equipment, good coffee and water and proper brewing methods are required to make good coffee

- Coffee flavors are best extracted at 205 F

- What was the temperature of the coffee in the infamous lawsuit where an elderly woman sued McDonald's because the drive-through coffee she spilled in her lap was "too hot"? (180 F)

- Before serving, wipe the spout or mouth of container with a serviette, warn guests of action and pour from 1-2 inches above the rim of the cup, placing the sugar and cream on the table immediately before or after pouring to the right of the cup or center of the table

- Beverages should be poured and cleared from the right with the right hand

- The cup handle should be at the 3:00 position with the teaspoon to the right of the cup pointing upward

B. Hot Chocolate

- Chocolate is derived from the bean of the semitropical cacao tree. The bean is roasted and broken into small pieces called nibs which are ground, producing bitter chocolate.

- Cocoa is produced after extracting some of the cacao butter (fat) from chocolate. Breakfast cocoa contains 22% or more of cacao butter. Hot chocolate is a blend of cocoa, sugar, milk and other ingredients

- Hot chocolate may be garnished with whipped cream, shredded chocolate and cinnamon.

- The service of hot chocolate is very close to that for coffee and tea.

C. Tea

- Tea is an aromatic beverage obtained by infusion of tea leaves with boiling water. The word originated from the Chinese T'e

- Tea has been widely consumed in the United States at an average rate of two pounds per year per person

- In processing, all teas have their leaves first rolled to release juices and develop flavor. The next steps vary according to the type of tea to be made

    - Green tea is not fermented but goes directly into drying after rolling
    - Oolong tea is lightly fermented and then dried
    - Black tea is fully fermented and then dried
    - The best tea comes from higher elevations

- Hot tea should be brewed in freshly boiled water above 185 F for up to ten minutes.

- Tea should be put into a metal or ceramic pot holding at least one pint and have boiling water poured over it

- After five minutes, the tea bag should be removed to prevent over extraction

- It is good service to bring the guest a pot of hot water with the tea on the side so that the guest can control the process

- Milk and lemon are customarily served with tea with the basic service procedures as for coffee

- Some operations offer a large variety of teas with added flavors (the text illustrates a box of teas being offered to the guests for their selection)

- Iced tea is usually made from orange pekoe black tea with double strength due to the dilution of ice

- Clouding is caused by making chilled strong tea or pouring it over ice as chilling precipitates' the tannis of the tea. This is why iced tea is newly brewed each day and is not refrigerated

- Iced tea is served in a tall glass with ice and a teaspoon. Sugar, brown sugar, and artificial sweetener should be brought to the table prior to the service of tea. Lemon is served with tea, preferably in wedges brought on a small doily lined plate

- In the south, iced tea is offered "sweet" or "unsweet" and is often referred to as the "house wine of the south". What tea service are students familiar with? (various)

VII. Hospitality Behind the Bar

- Excellent bar service relies on good organization, efficient control systems and talented, knowledgeable staff

- Bartenders should come to work in clean, neat uniforms, checking before the shift to ensure that everything is in order, neat and clean

- Inventory is usually taken of the storeroom before opening with items stored in their proper places and adequate supplies in place for the shift including liquor gun supplies and connection, bar bottles with pouring caps, and glassware

- Offering food when serving alcohol is good hospitality and an important part of responsible service as food generally slows the absorption of alcohol into the bloodstream. Fatty and high protein foods are best at this

- Mise en place is one key to good bar service when there is a rush

- House liquors or well stock are typically kept on lower shelves or in a well along the bar. Call stock are brands that guests ask for specifically rather than the well stock. They are often premium brands that come at a higher price than house liquors.

- A guest ordering Scotch and Soda is asking for _____ and a guest ordering Chivas and Soda is asking for____? (house or well and call respectively)

- When mixing in front of guest the bartender places the glasses on the pour and mixes the drinks in a container or pours from bottles then puts down bar napkin and slides the drink to the guest. Straws, stirrers and other garnishes should be placed as the drink is put in place. Except for rind twists, garnishes should not be touched but put in place with a toothpick or decorative sword.

- Liquor guns are not popular with many guests due to lack of flexibility in drink strength. Free pouring is often allowed but makes it difficult to control portions and costs. Jiggers can be used to measure consistent amounts

- Standardized recipes establish a strong foundation for customer satisfaction and profit

- How would standard recipes aid in responsible service? (Know how much guest is consuming)

VIII. Serving Spirits

A. Serving Straight Drinks

- Straight means the spirit is served alone without a mixer. If ice is served, the drink is "on the rocks". Spirits, wines and beers are usually served straight although they may be blended with other ingredients

- Straight spirits are typically served in a 7 ounce "old-fashion" glass but may be brought to the table in a smaller glass

- Some operations like to bring water or soda to the table with the drink, if used, it is no longer a straight drink

- Why would a mixer be brought when the guest orders the drink straight? (To dilute the alcohol in the bloodstream and encourage responsible consumption)

B. Serving Mixed Drinks and Cocktails

- Mixology refers to the art of mixing drinks. A mixed drink is any drink in which one beverage is mixed with one or more alcohol or nonalcoholic ingredients

- There are four basic mixing methods:

    1. Build- mix step by step in serving glass, adding each ingredient one at a time
    2. Stir- mix ingredients by stirring them with ice or in a mixing glass and then straining the drink into a chilled serving glass, minimizing dilution with ice
    3. Shake- mix by shaking with a hand shaker or by mixing on a shake, mechanized mixer
    4. Blend- mix in an electric blender. This must be done with any drink that contains solid food or ice

- Servers and bartenders should know the drinks usually requested by guests. If others are ordered, the bartender can look them up in a recipe book which should be kept at the bar

- What bartending recipe book is typically used? (Mr. Boston Deluxe Official Bartender's Guide, now in its 49th edition, first published in 1935. The book is published by a distiller and thus names the Mr. Boston brand in recipes but is still considered a good reference source at an affordable $2.95)

C. Serving Cordials/Liqueurs

- Liqueurs are usually served in small stemmed glasses that hold from 1-1 ½ ounces or with ice in a larger bowled glass called a brandy snifter
- Cordial carts may be brought to the guest by the captain or server enabling guests to select among the various brands on the cart
- Ask students to investigate the legality of the cordial cart in your municipality. Some states ban the use of these as they are thought to overly promote the sale and thus consumption of alcohol. (Various)

IX. Serving Wine

- Knowing how to serve wine can increase sales and guest satisfaction. Wine is served in food service and also bar operations that sell beverage alcohol
- Servers must have good product knowledge to sell and skills to serve requiring training.
- Many find chilling a dry or medium dry white wine increases its crispness but over chilling can result in a loss of bouquet and taste. White wine is best when served at 45-55 F and reds at about 70 F
- Marble sleeve chillers are now often used in place of ice buckets. When using an ice bucket, the ice should be mixed with water at 1:1, taking about 15 minutes to bring a bottle to 50 F. Bottles removed from the bucket should be wiped thoroughly. When guests are finished, glasses should be removed promptly.
- Some wines develop a sediment called lees and should be decanted. Sediment can be left in the bottle by gently opening a bottle then carefully, slowly pouring the wine into the decanter from which the wine will be served. A short burning candle under the neck will reveal the appearance of the sediment
- A sommelier (wine steward) is employed by some operations to give out the wine lists, take the order, bring glasses and serve but servers must do this in most operations
- The server should have sufficient knowledge to suggest wines based on the preferences of the guests, tastes, entrees ordered, and specific price range requested
- The pivot point system and established abbreviations should be used when taking orders. Guests ordering a bottle should be watched so that the server can pour for them, if desired. Be prompt in requesting an additional order when the bottle is empty or nearly so.
- There are many glasses, practically all stemmed, designed for holding wine. It is not necessary to of offer a large inventory of wine glasses if using an all purpose wine glass.

Champagne and other sparkling wines are more often served in fluted glass, aperitif or dessert glass

- Servers should note the importance of proper service to guests
- Servers should never make any adverse comment indicating that the guest is lacking in knowledge about wine (or anything else)
- The bottle is brought to the table and presented to the orderer unopened to check on the accuracy of the selection, held from the bottom and neck, at eye level, handling with care
- To open, rest the bottle on the right of the ordered on the table, cut foil leaving the bottom around the neck. Wipe the top neck of the bottle and cork and insert the opener into the center, remove with a clean, smooth pull.
- Bring the bottle to the right of the orderer and pour about an ounce so it can be tasted and approved (or not) The bottle should be twisted gently at the end of pouring to prevent dripping
- Remove the cork from the corkscrew and set it on a folded serviette or small plate to the right of the taster. Some will choose to sniff the cork or check for a dry cork indicating improper storage
- Serve the guests in a clockwise manner, ending with the orderer, some operations will serve women before men
- Fill glasses about 2/3 full, sparkling wines tend to bubble when poured, leaving a bit of foam
- After pouring, place the bottle of white or sparkling wine into a bucket or cooler at tableside. Rest red wine to the right of the orderer. Older reds are placed in a wine basket that holds the bottle on a slant so the lees are not disturbed.
- How has the shortage of corks and introduction of plastics or twist tops to finer wines changed the service of wine? (Will alter routine, servers need to check before opening to see what the situation is. It would seem ridiculous to place a twist top on a napkin or plate for the host/ess)

X. Serving Sparkling Wine

- It is important to know the correct method for opening a bottle of sparkling wine such as Champagne.
- The bottle must be chilled and not disturbed before opening to avoid the danger of a projectile cork

- Cut the foil around the top of the bottle
- Place hand on top of cork, leaving it there until cork is removed. Cut wire
- Wrap towel around bottle for safety and spillage- remove cork gently, slowing turning the bottle in one direction, cork the other, ease cork out
- Pour the sparkling wine into appropriate glassware

- What sort of occasions do guests order sparkling wine for? (Anniversaries, birthdays, New Year's, weddings, engagements, and other festive, happy occasions. Sparkling wine is associated with celebrations)

XI. Serving Beer

- Beer may be served in pitchers, mugs, and a variety of glasses or in a bottle accompanied by a glass. When serving, be sure to wipe the bottle before bringing it to the table

- Some guests prefer to pour their own. If the server pours, it should be carefully in the center of the glass so that an inch of foam appears, coming up to the rim of the glass. 10-12 ounce glasses are common.

- Serving beer from a can is not recommended but pouring procedure same as for a bottle with a quick movement of the wrist to stop drops from falling from bottle or can.

- Servers should serve draft beer within 30 seconds of its drawing. It is unappealing if served flat. The foam should be 1 ½ -2" high. Draft beer ordinarily comes from a keg kept refrigerated at about 40 F and is carbonated as it comes out of the spout.

- Draft beer is very perishable because it has not been pasteurized to destroy harmful organisms. Canned or bottled draft beer is filtered to remove these making it unnecessary to pasteurize the product which would destroy the draft taste.

- Are students familiar with other methods of service than those mentioned here? (Pouring beer down the side of the glass is thought to prevent overly foamy heads. Some may have been taught this method and used it effectively) Many premium beers may come in cans and the association with a "lesser" product may not be apt. Some Mexican beers are properly served with a slice of lime or lemon. Weiss beer is properly served with a lemon wedge.

XII. Serving Nonalcoholic Drinks

- Milk is usually served chilled and without ice. In more casual operations the carton may be brought to the table. Small children are often served in a plastic cup with lid and straw

- Milkshakes, malts and other mixtures are served in tall, thick glasses topped often with whipped cream and a cherry. A straw and ice cream spoon should be served with them

- Juices are served in a 6 ounce glass though 8-10 ounce glasses are also used. They should be chilled but seldom iced. If servers pour their own juice, care must be taken that liquids are properly handled and refrigerated.

- Bottled waters may be poured from the container and brought or the opened bottle may be brought and poured by the server. When serving it is proper to place the glass of ice on the table and pour by filling about a half of the glass or as guests requests.

- Carbonated beverages are often served throughout the meal. Some may be garnished. The service for carbonated beverages is the same as bottled waters. If served from a soda gun, they must be ordered from the bar.

- Guests ordering freshly squeezed juice, smoothies or ice cream drinks are sometimes given the pitcher that the drinks were blended in on the side to refill their glasses with what is "their" purchased beverage. This is served on a napkin to the right of the glass. Formerly, the remainder may have been simply thrown away. The newer presentation is thought to enhance value perception.

XIII. Serving Alcohol Responsibly

- There is a greater chance of avoiding legal and criminal action and successfully handling problems related to alcohol beverages if there is a responsibly planned and continuously reviewed set of policies and procedures for every aspect of service. The following are recommended;

    - Written policies and procedures include never serving to the point of intoxication, serving someone already intoxicated or serving minors

    - Have a good training program in alcohol service and provide disciplinary procedures for those who disobey policies and procedures

    - Eliminate promotions that encourage over consumption such as 2 for 1s and oversized drinks. (* These are illegal in many states)

    - Check every young guest's ID to ensure that minors are not served

    - Offer food and nonalcoholic beverages

    - Ensure that intoxicated guest does not drive away from the establishment. Call the police if one refuses assistance and does drive away.

    - Record all incidents in a log book

- Alcohol service is big business but one carrying more legal responsibility than in the past

- If a guest becomes intoxicated and is injured or injures another, the server and established can be charged with third-party liability with heavy fines and prison sentences possible

- Courses in alcohol awareness are required by many states. Training includes consequences of overserving, how to tell guests they can no longer be served, and what to do when someone arrives already intoxicated. Server must be trained to assess each situation

- Discuss the reasons for liability laws (Party that profits from sales of dangerous substance; weapons, explosives, poisons, alcohol, addictive drugs, bear a responsibility for their safe sale, "deep pockets" of business versus often uninsured guest who is irresponsible, political pressure to end drunk driving, etc.)

- Alcoholic beverage service laws differ from area to area. First rule is to know and obey laws applying to the establishment.

- Most states use the BAC (Blood Alcohol Concentration) test to determine intoxication. A BAC of .08 or more is considered intoxication in all states.

- The amount a guest can legally drink depends on the alcoholic content of the drink, the weight of the guest and the time that passes during consumption

- One system calls a guest with low BAC in the green zone, approaching intoxication in the yellow zone and in danger of or already is intoxicated in the red zone.

- Carefully check Ids. Many establishments do so for anyone appearing younger than 30. Make sure Ids are valid and not borrowed or faked. Bars or establishments patronized by young people often set up ID checking stations at each entrance and staff it with trained employees.

    - Ask for identification

    - Examine identification. Verify photo, date of birth, for questions, compare to ID guidebook and either obtain other form of ID, politely refuse service, or contact management to handle

    - Thank customer and return identification

    - Check everyone in party who is questionable

- State liquor codes govern liquor licenses. One should establish a good relationship with law enforcement and regulatory agents by adhering to code and practicing good business ethics.

- In most states it is illegal to serve alcohol to minors under 21 and people who are already intoxicated. There is no penalty for refusing to serve someone you suspect is a minor or intoxicated. You have the right to protect your guests, self and establishment's liquor license.

- Students often have strong views on this topic. Guided discussions can be very valuable. Remind students that no other industry has a higher incidence of alcohol and drug usage amongst employees than ours.

Chapter Review Questions Answer Key

1. Aging improves the quality of most spirits that are aged in barrels or wooden kegs. Spirit aging does not continue once it has been bottled. Wine is also improved with aging in barrels and casks. Wine does continue to age in the bottle. Beer should be consumed as soon as possible as aging does not improve it. Keg beer has not been pasteurized and must be kept at or below 40 F to avoid contamination.
2. BAC is blood alcohol content. Most states say that a level of .08 is legal intoxication. An ordinary person can break down one ounce of alcohol per hour.
3. If a spirit is 80 proof, its alcohol content is 40% by weight.
4. Bourbon is different than other types of whiskey as it must be made from at least 51% corn as fermentable material, normally aged four years in charred oak barrels at 125 proof. Kentucky bourbon must come from that state.
5. Fermented sugar cane or molasses is used to make rum.
6. Wine glasses should be filled half way for wine tasting and 2/3 full for still or sparkling wine service. With sparkling wine, there should be a pause for foam to subside if more filling is desired.
7. The qualifications for a bartender should include; alcohol knowledge, suggestive sales, personable/entertaining, responsible service skills- perhaps obtained through a certification training program, cleanliness and neatness, ability to perform inventory and account for cash drawer, and ethical
8. Well stock is house liquor. When a guest orders a rum and coke, for example, they are ordering house liquor. Call brand is often a premium brand ordered by the guest by name, Mount Gay and coke for example. The call brand is generally more expensive than the house and carries a higher price for the guest.
9. Check young guests' identification and politely refuse service to anyone who does not have identification, whose identification does not appear to be legitimate or who is underage. The server should also keep an eye on the party to ascertain that no one of age is ordering alcohol and allowing the minor to consume it.
10. Present the bottle to the guest, unopened so they can examine the label and ascertain that it is what they were expecting. Cut the foil from the neck of the bottle in a quick motion, leaving the bottom foil around the neck. Wipe the top of the bottle and the top part of the cork with a napkin or serviette. Insert the opener into the center of the cork and get a clean, smooth pull. Remove the cork.

11. The answers will vary but all but the Irish whiskey should have some smoky flavor.

12. The answers will vary. Each should have the flavor of their predominate agent; Sambuca- anise, Curacao- orange, Gin- neutral or botanical, Vodka- neutral or flavored, Rum- sweet, Cognac- heavy, Tequila- distinctive

13. These too will vary. The sparkling wine is recommended with hors d'oervres and desserts, Cabernet with steak and lasagna so that meat and cream might score well, the Chardonnay with roast pork, grilled or broiled fish and cream sauce, Merlot with beef or lamb, Pinot with seasoned or marinated meat.

14. Answers will again vary. Stout is recommended with desserts, as is porter, Weiss is a wheat beer with a yeasty flavor, Pilsner is said to go well with salads and pale ale with spicy food and soup.

15. Answers will vary.

* Note 11-15 call for a fair amount of alcohol consumption. Be mindful when making this assignment regarding the maturity level of students.

Chapter Cases Answer Key

The Ideal Server

Various answers can be given though they should encompass the principals of the text. Various duties of a server are described in the appendix, page 238 and might serve as a basis.

The Award Problem

The management has made a mistake and Tobias will certainly suffer morale issues if he is not rewarded. Future sales contests would certainly be doomed. To start out, painful as it is, the only fair way to resolve the current problem would be to give Joe $500.00 and consider it an expensive lesson. Future contests might begin with a base amount to be divided up with more if sales go above a set level. The dividing could be according to total sales with categories for goals in various categories such as; most overall sales, highest number, most improved, etc. with prizes for top categories beyond cash being gift certificates, first pick of days off for a month etc. Students' answers will be various but should go along the lines of sharing rewards so everyone believes that it is worthwhile to try. It is also important not to promote sales of a product the guest does not want or irresponsible service simply to win. This may happen if there is only one big prize.

Chapter 9 Test

True/False

T F 1. Far older than written history, are cave pictographs showing humans drinking beverages (presumed to be alcoholic) in celebration.

T F 2. Beer is by far the most popular alcoholic beverage and in 2003 made up more than 30% of all alcohol sales

T F 3. 100 proof spirit alcohol is 100% alcohol by weight.

T F 4. Cordials distinguishing characteristic are their smoky, heavy bodied flavor.

T F 5. Still wines are a result of allowing the C02 gas (carbon dioxide) to escape.

Multiple Choice

6. The most widely consumed beverage today is:
   a. Wine
   b. Beer
   c. Coffee
   d. Orange Juice

7. The blood alcohol content (BAC) associated with the legal level of intoxication, blurred vision and slurred speech is:
   a. .05
   b. .08
   c. .15
   d. .40

8. Any spirit distilled from a mash of fruit or fruit derivatives can qualify as:
   a. Brandy
   b. Liqueur
   c. Rum
   d. Wine

9. Which of the following is NOT one of the three types of wine categories?
   a. Table
   b. Still
   c. Sparkling
   d. Fortified

10. White wines can be made from:
    a. White grapes only
    b. White or green grapes only
    c. Pressed grape juice that has been fermented but that has not been aged
    d. Red, green or white grapes that have been separated from their skin

11. The color of wine that tends to be heartier, heavier and more flavorful than their counterparts tends to be:
    a. White wine
    b. Blush Wine
    c. Sparkling Wine
    d. Red Wine

12. The most commonly used grain in beer making is malt made from sprouted:
    a. Corn
    b. Wheat
    c. Peat
    d. Barley

13. The best way to store ground coffee in order to preserve its flavor is to:
    a. Pre-pour it into filters, stack and keep prepped for one week at a time usage per forecast so that it can be brewed immediately after being ordered
    b. Do not store, but pre-brew and chill coffee, boiling each pot to order
    c. Keep vacuum packaged or at freezing temperatures
    d. Keep it in a tightly closed container in dry storage

14. What sorts of bar snacks would work best at slowing the absorption of alcohol into the bloodstream thus assisting in responsible service?
    a. Chicken wings and pizza
    b. Raw vegetables and dip
    c. Rice crackers
    d. Fruit

15. In most areas it is illegal to serve alcoholic beverages to all except the following:
    a. Minors under 21
    b. Guests who are not able to show proper photo identification
    c. Guests who drove to the establishment and intend to drive away.
    d. Already intoxicated guests

Short Answer/Essay

16. Describe, in order, the process of tasting wine.

17. What are the three ways wines are named in the United States?

18. What process is involved in pairing wine and beer with food?

19. Describe how to properly serve beer to a guest who is of age and not intoxicated.

20. Describe what is meant by third party liability and why it is of concern in the restaurant and bar industry.

Chapter 9 Test Answer Key

True/False

1. True
2. False. Where beer is by far the most popular alcoholic beverage, it made up 50% of sales in 2003 not 30%.
3. False. 100 proof spirit alcohol contains 50% alcohol by weight not 100%.
4. False. Cordials are often distinguished by their sweetness.
5. True.

Multiple Choice

6. c
7. b
8. a
9. b
10. d
11. d
12. d
13. c
14. a
15. c

Short Answer/Essay

16. First, hold the wine up to the light to view the color, next, swirl and smell the wine. Do not swirl sparkling wine. Smelling should be done with the nose inside the glass. Tasting determines flavor, and other sensations. The impression left in the mouth is called the finish. This is the evaluation period of a wine tasting

17. In the United States wines are named by the predominant variety of grape used (varietal), by broad general type (generic) and by brand name. Imported wines may also be named by their place of origin (though this was not asked)

18. First, the food, wine and beer of the establishment should be tasted so that they can be described and recommendations can be made. There are guidelines, established so that the flavors are complimentary but it should always be remembered that the taste of the individual guest is the primary concern. Lighter foods generally are complimented by lighter flavored/bodied wines and beer and heavier, spicier or fattier foods by more bold flavors that can cut though fat.

19. Beer may be served in a pitcher, mug or variety of glasses. Be sure to wipe the bottle first and offer it to the guest with a chilled glass to determine of the guest wishes to pour their own or have the server do it for them. The guest's preference should be observed for each subsequent beer served. The beer should be poured carefully down the center of the glass so that the desired inch of foam appears as a collar around the top of the glass. The bottle should be given to the guest after the pour is finished if it is not empty. Can service is the same. Draft beer should be served within 30 seconds of its drawing. The foam should be 1½-2".

20. Third party liability refers to holding the server and operation responsible for a person who drinks and is served in the establishment and injures others because of the intoxication. Heavy fines and prison are possible. This evolved because of numerous court decisions that have increased the liability of those engaged in serving alcohol to the public.

Activity/Training

Have class perform role plays for the following situations:

- Checking the identification of a person who appears to be a minor and whose license does not seem to be authentic. The guest is with others who have legitimate identification and want to celebrate an occasion with a toast.

- Refusing service to an already intoxicated patron. The patron arrived at the establishment in their own vehicle and intends to leave with it if they are not detained in some manner.

- A guest wants to impress their table of friends and order the "right" bottle of wine. One guest is ordering lasagna, another grilled chicken, a third seafood and the fourth, veal. Have the server practice suggestive selling to the table.

- Perform, completely, the ritual of bringing wine to the table, opening the bottle and serving it to the guests either with a real bottle of wine or using a prop.

- Performing "last call" while practicing responsible service. The table wants to order more drinks than there are guests and/or desires "to go" containers.

Chapter 10: Management's Role in Service

Objectives

- Describe management functions necessary to a successful operation
- Explain how motivating and training servers helps an operation deliver excellent customer service

Outline

I. Introduction

- Managers establish service standards, motivate servers, schedule, and train and provide them with the equipment, tools and environment they need to do the job well.
- Good service is a partnership task between managers and servers working as a team
- Management must set high standards, communicate them and see that they are met or exceeded and support employees through training and recognition
- What would be wrong with each server setting their own standards? (Lack of consistency, may fail to meet guest expectations or be confusing)

II. Establishing Service Standards

- Standards are specific rules, principles or measures established to guide employees in performing their duties consistently
- Standards allow management to measure and evaluate employee performance and operation performance toward pleasing guests.
- Service standards include service policies and mechanics models and are based on the type of operation and quality management wants to achieve
- The first step in managing and delivering quality service is to set up and communicate service standards. This involves seven distinct steps:

    1. Set standards and describe in detail
    2. Establish policies and procedures for accomplishing standards. A policy is a plan a procedure is the manner the plan is implemented.
    3. Provide necessary space, equipment and environment to achieve standards

4. Provide training and guidance to ensure that standards are met.

5. Review with employees so they know and understand how standards performance will be measured.

6. Train employees to perform specific tasks to meet standards. Follow up with checklists, sidework, tasks lists and job descriptions

7. Encourage and seek out employee feedback so management can be aware of problems

- Establishing standards is relatively simple. The difficulty is in training employees and seeing that they follow them consistently

- Standards vary from fine dining to casual. The text discusses typical full service standards but each operation must interpret what will best suit their guests

- Ask how the provision of a napkin to guests might vary according to establishment (Fine dining, linen, folded per operation choice, casual may be paper, possibly from dispenser. In each situation, guests are provided with a supply in order to protect their clothing, and dab their mouth or fingers)

III. Management Functions

- Good service occurs when operations are well managed. A poorly run organization is unsettling to employees and leads to poor productivity and performance. The five management functions to see that an operation is properly run are:

A. Planning

  - Starts with establishing a mission and goals for the organization and its people.

  - All employees must have a stake in goals or they won't be realized. Employees must have input and realistic plans, subject to revision to follow.

B. Organizing

  - Applies to all processes and resources including people

  - An organizational chart shows an operation's positions, and their relationships to each other including who reports to whom.

  - Organizations where responsibility flows from the top down is known as a line organization

- Unity of command ensures that all employees report to only one supervisor

- Reverse Pyramid is an approach to replacing the line organization with a team effort. Manager is at the bottom of the pyramid, supervisors in the middle and front line employees are on top so they can better serve customers

- Is the reverse pyramid a good fit with the hospitality industry? Why or why not? (Driven by guest satisfaction is positive, some employees at line level are not invested- negative)

- Delegation is where responsibilities and tasks are delegated to employees, though the delegator does not escape responsibility for seeing that the job is done properly. The employee must be given the authority to accomplish the work.

    - Advantages to management of delegation:

        - Delegation helps manager perform better
        - It can establish pride in the employee charged with new responsibility
        - It can save managers time by freeing them for more important tasks

    - Proper delegation requires the following:

        - Delegatee must know how to do delegated task
        - Person delegated must be willing to do it
        - Person delegating must provide resources and assistance to do the job including granting authority
        - Ask for examples of tasks that a manager might properly delegate to a server and those that would be ill-advised (various)

C. Staffing

- Hiring qualified employees is the first step in creating an excellent service staff

- Person hiring must ascertain those seeking positions have necessary abilities and attitudes and reject those that do not

- One builds a staff by determining what tasks must be performed to achieve the desired goal and then allocating positions so that these tasks are performed

- Job specifications are then written up, indicating what the person in each position does

- The next step, is to hire people who can adequately fill these positions

- A staff consists of a group of individuals who are working together toward a common goal. A number of specialized positions are needed to perform various tasks

- In food service, there are culinary, server and managing staff and perhaps, others.

- Within each staff will be various positions, for example, within service staff are host/ess, servers, buspersons and others. Each work together as a team to perform its functions and the various staffs blend their efforts together to achieve their goal

- Individuals working alone, as opposed to as a team are without help. Team membership helps carry one easily and smoothly.

- Using the job description for a server, exhibit 10.3; list the skills you would seek out in an applicant. How would you determine if a person possessed these skills? (Technical skills or ability to perform; tray handling, organization, knowledge of food, human skills; warmth, helpfulness, communicative, problem solving, courtesy, mathematical skills etc.) (Ask them to demonstrate tray handling, role play resolution of complaint, ask them to describe the funniest thing that ever happened to them, basic math calculations, etc.)

D. Leading

- An operation cannot succeed without effective leadership

- Good leaders must know what the goal is and take firm and adequate steps to reach the goal, command respect, good will as well as success and profits.

- Good leadership motivates workers to do their best, sees they are treated fairly and given a chance to reach their goals.

- The recommended leadership style for today is participative leadership where the manager is part of a team working with the employees toward a common goal

- The authoritative or dictatorial leadership styles traditionally used in hospitality allowed little or no employee input and used fear of discharge or punishment to motivate.

- Participative leaders act as coaches to lead teams to success. They seek out ideas and opinions, and often follow them giving employees a feeling of value and respect

- Participative leaders still must ensure that standards are met and rules are consistently and fairly applied

- Managers should correct words, actions or attitudes, not the employee personally. Counseling sessions should be held in private without any tones of anger or threats. Employees must feel that they are being offered help rather than criticized.

- Have students' role play the same situation (guest walks out without paying-"dine and dash") with an authoritarian leader and a participative one.

E. Controlling

- Controls lead an operation toward achieving its goals. Without them, operations lose money and workers become frustrated and unhappy.

- Establishing standards and procedures for meeting them is a key element of controls.

- Servers and managers should check continuously to see that deviations from standards are avoided or corrected.

- Wouldn't workers prefer there to be no rules? Why would they become frustrated and unhappy without them? (Chaotic, complaining guests, loss of income, etc.)

IV. Motivating Servers

- Management is responsible for seeing that all servers are motivated and that they work in a motivating environment. It is a challenging task that should be one of management's most important goals.

- People are motivated to work by a number of factors:

    - Basic needs for food, housing, transportation etc.
    - Safety needs for secure work environment
    - Social belongingness and change of environment
    - Self and ego gratification
    - Self actualization

- Not everyone works for the same reasons. Managers need to seek to satisfy all and seek to extend them to higher levels. The result is more satisfied employees.

- Ask students what their aspirations are with the industry and what motivates them now. (various)

A. Hiring Motivated Servers

- Motivated employees can often be detected in the interview by exhibiting the following personality traits:

    - Genuine interest and respect for people
    - Warm, outgoing attitude
    - Sincere wish for the job
    - Habits that will not interfere with the job
    - A willingness to accept an entry level position if inexperienced
    - A positive work record
    - Poise and confidence
    - Ability to solve problems and make good decisions
    - Pleasing appearance
    - Experience working in good operations, if applicable
    - Positive reasons for working other than income
    - Willingness to grow in the job; promotable
    - Good self-direction and self-reliance

- People must possess self motivation to work hard and serve enthusiastically. The following questions might help in evaluating motivation level or potential:

    - What was the biggest work challenge and how did you handle it?
    - What are your strengths?/Weaknesses?
    - Had problems following instructions?
    - What type of goals have you set for self?
    - What do you see yourself doing in five years?
    - Do you perform server duties with enthusiasm and dedication?
    - Do you like people?
    - What do you expect from coworkers?
    - What do you expect from your supervisor?
    - What new idea have you introduced that improved yours or others work and how did it make you feel?
    - How would you handle an irate guest due to a late, cold order?

- Have students pair up and ask each other the questions.

- Illegal questions that are not job related include:

    - Have you ever been arrested or accused of stealing
    - Do you have a green card?
    - How long have you lived at this address?
    - Are you the wage earner of your family?
    - Are you married? What does your spouse do?
    - How many children do you have/plan to have?

- How old are you?
- What is your credit rating?

- Ask students why these questions are not allowed. Many students will believe some ought to be. Ask for acceptable question or reason:

1. Have you ever been convicted of a crime?
2. Are you legally able to work?
3. Not allowed, does not pertain to job
4. Not allowed, does not pertain to job
5. Not allowed, does not pertain to job
6. Not allowed, does not pertain to job
7. Are you over 21 and legally able to serve alcohol?
8. Not allowed, does not pertain to job

B. Creating a Motivating Environment

- One of the most important things managers must do to motivate employees is to see that a respectful, pleasant atmosphere prevails. This helps servers feel secure and appreciated.

- A positive environment fosters team work and cooperation among servers who work in an often pressure-filled, fast paced and demanding job

- Creating a feeling of respect and commitment among team members assists the manager in ensuring a staff who want to do the best possible job consistently

- Good motivators include:

    - Establishing standards
    - Enhancing jobs through empowerment
    - Making the work easier
    - Encouraging open communication
    - Including servers in goal setting
    - Treating employees fairly and with respect
    - Creating awards for good performance

- Motivation destroyers include:

    - Assigning extra work without adequate compensation
    - Unfairness, aggression, abuse, coldness, inconsistency or standoffish behavior
    - Failing to provide good communication, supplies, tools or other standard conditions
    - Ridiculing or using sarcasm instead of constructive criticism
    - Over supervising and showing a lack of trust

- Failing to follow up on problems brought up by servers
- Making decisions without consulting servers that might interfere with their being able to adequately do their jobs
- Having too many bosses

- Ask students for their experience with motivators and demotivators. (various)

C. Empowering Employees

- Empowerment is a management tool in which employees are given the power to make decisions that achieve higher standards of service.

- This includes the authority and responsibility to make things happen such as the correction of problems without first needing manager approval.

- Guidelines may be provided to assist the servers in making appropriate decisions

- Servers may feel more personal commitment to serving guests well.

- Employers benefit as problems are solved more quickly and guests are better served.

- Discuss with class the sort of guidelines that might be developed for comping guests when problems occur (various- idea is to make up for the problem, not hit the jackpot. Preserving relationship and generating positive word of mouth is the goal.)

D. Rewards as Motivators

- There are many ways to motivate employees toward doing an even better job:

  - Positive feedback and recognition of a job well done
  - Monetary rewards and prizes
  - Employee of the month and year awards
  - Bonuses for longevity
  - Vacation days, dining credits, tickets, passes and trips
  - Achievement is traditionally rewarded with advancement
  - What can be done in small operations where there are no spaces to promote outstanding employees to? (Recommend for positions elsewhere, tuition reimbursement to Hotel-Restaurant or Culinary programs etc.)
  - Rewards should be open to all, with complete information, and evaluation of effectiveness. Reward should be given ASAP, explaining why winner qualified.
  - The reward should be worthwhile and the reward program respected

  - Benefits such as health care, educational assistance, EAPs, and even day care can be good motivators
  - Whatever the incentive, managers should give them consistently and tie to concrete behaviors- favors will undermine the plan
  - Respect and loyalty cannot be purchased; it must be earned through sincere actions. Manipulation will be seen through
  - Money alone rarely satisfies employees

- An employee survey found the following to be motivating; job security, good working conditions, appreciation, job satisfaction, good wages, acceptance, dignity and respect and good benefits

- Only two of the above (wages and benefits) are monetary, the rest cost little or nothing

- Unless servers get a feeling of gratification from doing a good job, they lose interest. Work must be meaningful to them and help them reach their personal goals

- Managers cannot reach their goals unless the employees reach theirs

- Ask class members to reflect on what their goals are and how their current employers might reasonably assist them to achieve those goals. (various)

E. Evaluations as Motivators

- The performance evaluation or appraisal is the formal method for providing feedback

- Management should make it clear that the purpose is to help employees develop and improve

- Evaluations should let employees know their opportunities for advancement

- Evaluator should look at customer comments, dollars produced, tips, covered served, and other factors as well as intangibles such as; attitude, teamwork and punctuality

- Appraisal information can be important in giving advancement, pay raises, rewards and what management can do to help employee further their goals

- A good appraisal leaves employees more sure about how well they are doing and what they need to do to improve

- Ask classmates if they would like to share some of their experiences with appraisals and discuss (Expect that many will not have received appraisals

V. Scheduling Servers

- Proper scheduling can be a good motivating force
- It is essential that all workers be treated fairly in scheduling
- Station rotation and requests should be considered in light of fairness
- Discuss class members experience with having work schedules fit with school schedules. (Various)
- Managers are responsible for seeing that adequate, competent servers are scheduled at the right time. Too few make service poor, too many slows servers down and creates problems.
- Team service allows one server per station to attend to guests while others do other work. Buspersons help.
- Computer software programs can help with scheduling, station assignment, analyzing payroll and monitoring service
- Ask class members for experience with scheduled openers and closers. (Scheduling closers can help ease the tension that may be caused with the question of who is first to be phased out.)

VI. Dining Room Arrangement

- The dining area must be planned properly for good service
    - Aisles must be wide enough to allow for safe and good passage for servers carrying trays
    - Too many tables cuts down on server space and assures poor service
    - 12 square feet per guest is required for table service and 20 square feet per guest for counter service
    - Club and luxury dining areas often have more space, banquets require less
    - Distance between tables should be 4-5 feet with 24 linear inches per cover on the table, thus a table for four should be 30 inches square.
- Managers should see that there is adequate room for tray stands, service stations and other service equipment. One to four servers typically use one service station.
- If possible, water should be piped to the service station. In fine dining a permanent carving station may be set up or mobile stations used to offer flexibility.
- Tray stands, service stations, and other things needed to give good service should be kept in order and be sufficient in number.

- If the class is conducted in conjunction with a lab, have class try out various arrangements of the dining room, making up charts for those deemed most useful

VII. Kitchen Arrangement

- A smooth flowing kitchen arrangement assists servers in providing good service
- A smooth, one way flow with pick up arranged in sequence of courses is best though some stations may be responsible for multiple courses and/or multiple stations may be responsible for the same course
- Mobile equipment can be used, reducing server travel
- It is essential that food be delivered to guests with the proper appearance and temperature. Guests should not have to wait long for food.
- If class is conducted in conjunction with a lab, have class draw traffic flow in kitchen and make suggestions to improve the present system, if not, they can draw up their own or familiar operations' kitchens

VIII. Training Servers

- Servers must be trained in what to do, how it is to be done and the professional standard expected
- Managers are responsible for seeing that training is done by themselves or is delegated
- Though training can be costly, poor service is more costly in terms of lost guests and profits. Benefits of good training include:
    - Improved service and increased number of satisfied guests
    - Improved productivity, reducing turnover and labor costs
    - Reduced waste, accidents, and breakage
    - Opportunities for skilled servers to grow and advance
    - Lower frustrations (reduced turnover)
- Training should be ongoing for all operations, as one never stops learning.
- Training should improve knowledge, skills and attitudes

A. The Effective Trainer

- A good trainer knows people; how to teach and motivate them to learn
- The trainer must know the material and have a lesson plan

- Trainers must be good leaders who can control the learning session, adept at building trust and confidence

- The trainer is a listener as well as a speaker, stimulating student questions and involvement

- A student understanding the "why" of training is more likely to make the learning permanent than simply the what, when and how

- Repetition reinforces learning.

- Not everyone can teach- to some it comes naturally, though techniques may be improved

- The trainer should plan detailed training and orientation sessions. Pretests help identify which employees need the most training. Posttests and periodic retests ensures application on the job

- Have students list the aspects of the best class(es) they have experienced and discuss how those positives could be applied to training employees (various)

B. The Training Session

- Lecturing is good when the material to be covered is short and not too technical. Lecture sessions should be short and used with other teaching techniques

- Trainees remember: 15% of what they hear, 30% of what they see, 50% of what they see and her and 85% of what they practice

- Group discussions provide motivation, interest and subject retention. The trainer must identify the objective, relate to learning and manage the discussion, staying out of actually participating

- Role play allows servers to see and practice what is being learned

    - The first step is to give the trainee a situation relating to their position and allow the trainee to work through it.
    - Best used where active, physical effort is needed, coupled with the demonstration of some skill and knowledge
    - The trainer explains the proper way and students perform followed by a critique

- Show and tell gives students a chance to hear and see how something is done, perform it and see where improvement is needed. It is usually used in classroom settings. The following steps are used:

    - Tell the server how to do the task
    - Show what is to be done, how and why
    - Have the server repeat, verbally explaining what is to be done an d why
    - Have trainer and/or class review performance

- Observation and on the job training (OJT) usually occurs in the workplace under real conditions and can be quite successful

- Observation allows the employee to follow the trainer and, under a real situation, go through the same tasks, receiving feedback in private

- OTJ must not diminish service to guests- the trainer must not teach the wrong techniques

- Have a common small training task such as serving a cup of coffee and have class group members try using all the above training methods to teach it and see which is best. Next, try something like complaint handling and see if the outcome is different. (Different methods work for different training tasks)

C. The Training Space, Tools and Equipment

- Some operations have wisely established their own training manuals

- Job descriptions should indicate servers tasks and responsibilities

- Externally produced training booklets, videos, software packages, CD-ROMs and posters are available from professional associations and publishers

- Training models and formats vary according to size and type of operation.

- Ask students to bring in or describe methods used at their places of employment.

IX. Reservations

- Management is responsible for establishing the reservation system and seeing that it operates as desired

- Reservation taker should show warmth and cordiality in taking the reservation as this is the first impression of the operation

- The person should be greeted, questioned politely and the reservation book should be consulted before confirming; noting the date, time and party number, thanking the caller at the end.

- No-shows can create problems. It is advisable to ask guests to inform operation of change of plans and to obtain a contact number, particularly for large parties. The number can be called the day of the party to confirm

- Some operations take a credit card number and charge for no shows

- Ask class members what they think about this policy. Why is it sometimes needed? Why are reservations offered at all? (Guests sometimes make reservations at a number of places and decide at the last where to go or change plans and do not feel it necessary to inform the operation. Reservations are offered because guests want them and operations can better forecast based on them. Some operations only take reservations for large parties and run on first come, first served basis for others)

- Seating preferences should be noted, especially if smoking/non-smoking if the restaurant has a smoking section. Other notes are; occasion, and special requests.

- Reservations should be arranged to get the best turnover possible, staggering them so everyone doesn't come in at the same time

- Some operations do not take reservations at all

- To establish a reservation system, determine the number of tables and seats available and if tables will have to be moved for different sized groups. From this, set up a reservation chart

- Usually a table is filled at dinner for 1 ½ hours for one or two and longer for larger parties- breakfast is faster, lunch a little longer, meetings still longer. An understanding of how long tables will be filled helps the reservation taker with timing

- It is not unusual to overbook by about 10% due to no-shows, cancellations and slow walk-ins but it can be problematic when there are more guests with reservations than there are tables. The following is suggested:

    - Be consistent in handling situations
    - Ask the party to please wait, perhaps fibbing that a table has overstayed- if the wait is long, invite the guests to the lounge to drink a beverage on the house or promise a bottle of wine upon seating
    - Give the time of wait, if possible, being as accurate as possible. If unable to quote a time, tell the party their table will be ready as soon as the seated party leaves
    - When approaching several waiting groups, tell those still waiting that they have not been forgotten
    - A seat card containing information for the server can be a helpful tool and contain wait times, notes, names etc.

- Manager must work to see that the whole operation is one cohesive family, leading to more motivated employees and profit for the operation

Chapter Review Questions Answer Key

1. Standards are important to good service because they provide specific standards for the type of operation to meet the needs of the guests. Management must establish standards and communicate them to employees, training them on the policies and procedures needed to meet them.

2. Line organization is when responsibility and authority flow from the top down. It is the traditional method. Many organizations are replacing line organization with an approach called team effort. One version of this is the reverse pyramid which places managers on the bottom and front line employees and customers on the top.

3. In participative leadership, the manager is part of a team working with the employees toward a common goal. Rather than reprimanding employees for problems like the authoritarian leaders of old, they point out problems and help correct them. Standards tell employees what is expected. Leaders work alongside employees when necessary, creating a feeling of fellowship and the sharing of responsibilities.

4. Questions to ask a person applying for a job that do not violate their privacy or the law are as follows:
   1. What was the biggest work challenge and how did you handle it?
   2. What are your strengths?/Weaknesses?
   3. Had problems following instructions?
   4. What type of goals have you set for self?
   5. What do you see yourself doing in five years?
   6. Do you perform server duties with enthusiasm and dedication?
   7. Do you like people?
   8. What do you expect from coworkers?
   9. What do you expect from your supervisor?
   10. What new idea have you introduced that improved yours or others work and how did it make you feel?
   11. How would you handle an irate guest due to a late, cold order?

5. Faults in dining room arrangements that can cause problems for servers include: too narrow aisles for heavy tray carrying caused by putting tables too close together and having too little table space for diners. There also needs to be adequate space for tray stands, service stations and other service equipment. Equipment and supplies should be kept in order and of sufficient number for service. Kitchen arrangement faults include a backtracking or multi directional traffic pattern or lack of one and/or illogical pick up system with servers needing to travel so far as to not keep hot food hot, cold food cold and to allow the appearance of food to deteriorate, as well as sanitary standards.

6. Benefits of a good server training program include: improved service, improved productivity, reduced waste, accidents and breakage, opportunities for advancement and

lowered frustrations. These benefits will result in increased guest satisfaction, reduced costs, and reduced turnover resulting in increased profits.

7. When taking a reservation, the following information should be obtained from the guest: the name of the party, the date desired, number in party, time requested, special requests or information and contact number.

8. If two parties arrive at the same time with reservations and only one table is available, the host/ess should politely explain the situation and offer the guests the opportunity to communicate any special need to be seated (time sensitive engagement of any sort following dinner, diabetic guest needing to eat etc.) Just as with the airlines, the "volunteer" party can be offered complimentary beverages and whatever else it seems to take to make certain that all are satisfied with the arrangement. Of course apologies that a table set aside for the reservation is still occupied should be offered along with the sincere regret for any inconvenience caused by the delayed seating.

9. When a busy server is unable to approach a newly seated table in a timely fashion, other servers can greet the table, explain that the assigned server will be right along and go ahead and take the beverage order, explain specials or otherwise let the party know that they will be taken care of shortly. Management might consider letting another server take the table and then, when the busy server is available, assigning them the next seated table.

10. The duties that management must perform to ensure good service include; setting service standards, communicating them and training employees in the policies and procedures to meet the standards, hiring servers that are able to be motivated, motivating current servers, establishing a service team where servers provide input, are treated with respect and develop higher levels of rapport and motivation, and set up the dining room and kitchen so that the flow of service is smooth and the necessary equipment and supplies are available.

Chapter Cases Answer Key

Meeting Competition

Ralph's main problems seem to be:

- Competition has eroded his successful family restaurant business
- He wants his business to stand out from the competition
- His menu choices have not kept pace with market demands for nutrition minded options
- His décor is outdated and not as nice as another competitor
- He has failed to utilize techniques such as early bird specials to bring in business

Ralph is still succeeding and would like to retire a success. He is not ready to retire now and ought to use the skills and assets he has to stand and fight.

Ralph's union staff has a stake in the success of his business. He should call them together and let them know that he wants to revitalize and ask for input.

Ralph should carefully consider the ideas of the staff and also consider the following:

- Analyze his menu; preserve popular signature items and add health conscious choices in keeping with his successful theme
- Consider a décor facelift versus expensive total renovation

- Consider marketing options such as Early Birds, and other methods of bringing in business and advertise the changes, inviting guests new and old to try him again

Projecting Costs for a New Operation

Jonathan has two major cost categories that are a bit out of line with averages. They are Food and Payroll costs. He can look carefully at those categories for ways to reduce costs. If he believes that these are fairly set, he can look at ways of increasing, realistically his sales projections; if the family income is indeed, above average, perhaps his prices could be eased upward, raising income and lowering food and labor cost percentages. He could also see if sales income was too conservative of an estimate.

Students will have various specific suggestions that should be evaluated on the basis of realistic probability of success.

Hiring

The interviewer should first be given the guideline below and then given specific questions to ask such as those that were used in question #4.

- Illegal questions that are not job related include:
    - Have you ever been arrested or accused of stealing
    - Do you have a green card?
    - How long have you lived at this address?
    - Are you the wage earner of your family?
    - Are you married? What does your spouse do?
    - How many children do you have/plan to have?
    - How old are you?
    - What is your credit rating?

A student representative on campus might work on job fair days etc. Antonia also might consider being a guest speaker at career classes; inquire into co-op or intern students, offer tuition reimbursement or at the very least, work with students' schedules. Cash bonuses for current employees who bring in new workers certainly couldn't hurt.

Chapter 10 Test

True/False

T F 1. The responsibility for good service rests squarely on the shoulders of the servers.

T F 2. In order for delegation to be successful, the employee who has had a task delegated to them must be given the authority to accomplish the work.

T F 3. Everyone is motivated to work for money.

T F 4. Over supervising and making decisions affecting employees without their input is just part of being in charge. A strong leader such as this provides motivation for the team.

T F 5. Empowerment is the act of assigning extra work which is so motivating that servers do not need any form of compensation.

Multiple Choice

6. The first step in managing and delivering quality service is to:
   a. Determine how much the guest is willing to pay
   b. Hire good servers
   c. Set up and communicate service standards
   d. Have the best food in town

7. The act of establishing a mission and goals, done with input from all employees describes the management function of:
   a. Planning
   b. Organizing
   c. Leading
   d. Controlling

8. An organization in which management is at the bottom, supervisors in the middle, and front line employees are on top so they can better serve customers is called:
   a. Line organization
   b. Anarchy
   c. Reverse Pyramid
   d. Unity of command

9. Which of the following is an inappropriate or illegal question to ask of interviewees?
   a. What type of goals have you set for yourself?
   b. Have you ever been arrested or accused of stealing?
   c. Have you ever had problems following instructions?
   d. What was your biggest challenge in your current last job and how did you handle it?

10. Which of the following is NOT an example of a decision an empowered server should be allowed to make once they have been given guidelines?:

a. Provide a complimentary appetizer for a guest who has waited a long time to be seated despite having a reservation
b. Serve alcohol to the son of a regular customer. The guest may be underage but does not have an ID but the customer is a big spender and tips lavishly.
c. Select new server uniforms from several choices approved by management
d. Accept a check from a regular customer who left their credit cards in their other wallet, despite the fact that the operation typically discourages payment by check.

11. The purpose of performance evaluations is to:
    a. Squash unrealistic hopes for raises or advancement
    b. Help employees develop and improve
    c. Point out the problems caused by the employees' failure to perform and reinforce the consequences of not straightening up
    d. Weed out the weak and support the strong

12. Scheduling more servers than needed:
    a. Assures that if someone doesn't show up, the shift is covered
    b. Assures that excellent service is provided and is good practice
    c. Slows everyone down and ultimately, creates problems
    d. Allows management to send home those in rumpled uniforms, latecomers and those with a bad attitude.

13. When dining tables are placed 2-3 feet apart:
    a. Management makes more money as more diners may be seated
    b. Management makes less money as tables are further apart than they could be
    c. It is possible for diners to eavesdrop on others, adding to the entertainment and enjoyment of all
    d. It is very difficult for servers to carry trays safely and provide good service

14. Advantages of training include all EXCEPT the following:
    a. Opportunities to advance
    b. Reduced waste, accidents and breakage
    c. Reduced turnover
    d. It is inexpensive

15. The kind of training that occurs in the workplace under real conditions that can be quite successful is:
    a. On the job training
    b. Group discussion
    c. Role Play
    d. Training videos

Short Answer/Essay

16. Would it be preferable to have servers view themselves as independent contractors or members of a team in terms of providing excellent service? Explain.

17. Compare and contrast authoritarian leadership style with participative leadership style.

18. What are some of the personality traits that may indicate that a potential server is motivated in the interview?

19. What are some of the distinguishing characteristics of motivators and motivation destroyers? What do they have to do with management style?

20. How should the host/ess handle a situation whereby a guest with reservations must wait to be seated?

Chapter 10 Test Answer Key

True/False

1. False. Good service comes not only from servers but also from managers who must establish service standards, motivate servers, schedule, and train and provide equipment, tools and an environment to do their job well.
2. True
3. False. People are motivated for a variety of reasons. Providing for the basics, safety and security, social belonging, ego gratification and self-actualization are the broad categories of reasons. Managers must seek to satisfy each employees needs and urge them to seek higher levels to motivate. Money is only one thing that motivates people to work.
4. False. Over supervising and making decisions without allowing input of those affected by them are ways managers destroy motivation.
5. False. Empowerment is giving servers the authority and responsibility to make things happen through guidelines. Giving extra work without any form of compensation is not the definition of empowerment and is a motivation destroyer.

Multiple Choice

6. c
7. a.
8. c
9. b
10. b
11. b
12. c
13. d
14. d
15. a

Short Answer/Essay

16. Team members support each other in the provision of good service and allow blended efforts to achieve a goal. If everyone views themselves as working alone, competition, and a lack of cooperation is the result. The guest is not well served by an "it's not my station" attitude and senses an atmosphere of disharmony.
17. Both authoritarian and participative leadership styles have been used in hospitality management. Authoritarian was used in the past, but due to changes in the labor market and the industry, managers had to change their leadership styles in order to survive. While authoritarian leader's rules with little or no input, and used fear to motivate, participative leaders act as coaches to lead teams to success. Participative leaders seek out ideas and opinions of workers giving them a feeling of value and respect and enhancing performance.
18. A genuine interest and respect for people, warm, outgoing, sincerely wishing for the job, habits that will not interfere with the job, if inexperienced- willing to accept entry level job, positive job record, poise, confidence, problem solving and decision making abilities, pleasing appearance, experience working in good operations if applicable, positive reasons

for working other than money, willingness to grow in the job, promotable, and good self-direction and self-reliance.

19. Motivators are things that build esteem in individuals and responsibility in individuals and teams. Persons who know they are valued are more likely to want to perform valuable work. Motivation destroyers assume that employees are lazy and dishonest and must be watched every minute. The needs of the employees to do the job and notion that they might be able to enhance operational performance with their input are discounted. The participative management style works well with the motivation techniques described in the text while the authoritarian style dovetails with the motivation destroyers.

20. Be consistent in handling all waits. Ask the party to wait, if the time looks to be extended, invite them to the lounge for a complimentary beverage or relay an offer of wine when seated. Give the wait time as accurately as possible, or "when the party seated at the table departs". When approaching several groups that have been waiting, assure those still waiting that they have not been forgotten. A seat card may be used to give the server follow-up information to better serve the guest

Activity/Training

Break students into small groups and assign them the following areas as training topics: Serving wine, napkin folds, crumbing a table, flambé or Caesar salad preparation, refusing alcohol to a customer entering the "red" zone, suggestive dessert sales, and handling a guest complaint.

Have each group develop a training plan and prepare a group training session in their topic for the rest of the class. If the class is associated with a lab, an alternative assignment would be to give groups sections of an operational manual for the class to write and present on a due date. The manual can then serve as a basis for future classes. Alternatively, a reservation system could be developed, or restaurants could be visited with a critique of service done and a report made to management (for class purposes) as to how to correct any observed problems.

Chapter 11: Table Etiquette

Objective

- Describe the etiquette rules concerning special foods

Outline

I. Introduction

- The way people are served food and drink often influences how they eat and drink it
- Product placement at the table and items for their consumption must all be correct
- Servers should know the basics of serving and dining etiquette so guests can enjoy their meal
- "To possess good manners is to not tell others what good manners are" means that a server must never cause a guest to feel uncomfortable by assisting them throughout the meal
- Management must train servers in etiquette and service so that they can assist the guests comfort, not inhibit it.
- Is etiquette commonly taught to servers these days? Is it commonly taught at home by parents? (Society seems to be becoming more ill-mannered and may not be as aware of the rules of decorum and public social behavior as previous eras)

II. A History of Table Etiquette

- Rules of decorum arose to suit the needs of the culture. The first written record of these rules is near the end of the Middle Ages; 1474 in "Health and enjoyment within Decency" and covered the art of good living and manners
- "The Courtier" was a 1507 manual dealing largely with proper conduct in public, though it also covered dining
- "Galateo" was the most popular work, appearing in 1555; a short, unpretentious work that became the authority on good manners through the technique of the conversation between an old wise man and his nephew
- Other early notable books on table etiquette include: A Treatise on Manners by Erasmus in 1526, Robert de Blois's collection of principles, and the 1765 Antoine Le Courtin

work, Traite de Civilite summarizing all of the previous writing and adding advice of his own. Japanese, Chinese and Arabic literature also contain interesting accounts of good behavior when eating according to the customs of those cultures.

- Ask students to research modern books and other sources of information regarding table manners. (http://www.mannersinternational.com/etiquette_tips_table.asp, http://www.soyouwanna.com/site/syws/tablemanners/tablemanners.html, http://en.wikipedia.org/wiki/Table_manners, http://www.cuisinenet.com/digest/custom/etiquette/manners_intro.shtml, http://www.cuisinenet.com/glossary/tableman.html)

III. Principles of Public Dining Etiquette

- Rules for dining etiquette grew out of what is practical, makes for good social functioning, considering others, accommodating people in social situations, and facilitates the art of eating and dining

- They were not formed by one authority but evolved from common usage of dining practices and social relations. Authorities have collected, summarized and published them but people themselves formed them.

- Rules of etiquette are not set but change constantly as society changes. Today's rules are less formal than the Victorian era.

- Customs and etiquette differ from culture to culture and for different occasions, places and company.

- Servers should learn rules for eating out in public. They should be followed when appropriate with the realization that all rules depend on the situation at hand.

- Have students codify a pamphlet describing the etiquette of ordering, receiving, serving and partaking of pizza delivery products. (Various, should include telephone etiquette, providing a lighted clear way for driver, receiving, paying for and tipping delivery driver, informal sharing of food with appropriate accompaniments; napkins, condiments, beverages etc.)

A. Pre-Service Etiquette

  - When taking reservations, the employee should get the party's name, request for smoking/non- if applicable, number in party, arrival time and date and any special requests

  - If there is a wait, say how long and where the party should wait. Some operations won't seat a party until all members are present, others hold reservation for only 15 minutes, some call to remind on day of reservation, others charge credit card when party is a no show but they must be informed of this policy, coats should be checked so table and seats are clear

- Maitre d' or host should ask for guest seating preferences before leading them to a table. If that type of table is unavailable, guests should be informed of the estimated wait time, options etc.

- When guests are seated, server may place napkins on the guests' laps or allow them to do this themselves. It is considered improper for the guest to tuck the napkin under their chin or in clothing. Do not shake the napkin violently to open. Napkins can be placed on the table so guests can pick them up and unfold them

- One of the first things the server will do is to ask the guest for their drink order. It is impolite to urge the guest to drink an alcoholic drink if they prefer not to. Suggest a nonalcoholic beverage.

- Servers should announce specials and their prices slowly and clearly

- Guests who wish to share should have the server have this done in the kitchen and avoid doing it at the table. Tasting is done by the guest who has the food placing a small amount onto a passed plate or passing their own plate to the guest wishing to taste.

- Take-home containers are acceptable except at formal meals and banquets.

- Guests should raise one hand or call "server" in a soft voice to get the attention of their waitperson. It is considered rude to whistle, clap, snap fingers, strike a glass or cup, stamp0 feet or make other noises to attract a server.

- If a guest is served something they avoid eating, they should not mention it but simply not eat the offending item. If something must be removed from the mouth, it van be brought to the front and put onto a fork or spoon. Fingers are acceptable for such tasks as removing fish bones. Inedible items may be unobtrusively put onto the plate.

- Servers must cover up large spills with a napkin

- Guests should not reach but should ask to have things passed. Things can be taken from a dish using the Russian technique using the spoon and fork to lift and carry the item to the plate. Items passed with a handle should have the handle facing the guest receiving the item

- Items with lids such as butter, jellies and jams should have the lids removed, desired quantities taken out and the lids replaced. Servers should remove and replace empty items as quickly as possible.

- It is proper to wait until all have been served to begin eating but in large groups, some would have cold food. One should eat when served, in this case and it is polite for those who are waiting to urge those served to begin. Servers should serve all guests in a party at the same time.

- Servers should not remove dishes until all are finished however if someone is uncomfortable with soiled dishes in front of them the server should ask if they would like their plate removed.

- Ask students to discuss their own experience with feeling awkward about how to behave in social dining or observing others who were ill at ease. Did the server help or hinder the situation? (Various)

B. Tableware

- Especially in formal dining, eating utensils are given precise placement on the table for specific uses.

- Knives and spoons are placed to the right of the guest while forks are placed to the left.

- In the US, people cut their food then transfer the fork from the left hand to the right, a process known as the zig zag method, in order to bring the food to their mouths. Europeans do not use the zig zag method but use their left hand to bring food to the mouth.

- The fork should be held at a 40 degree angle with the tines pointing down, holding the food to be cut. IT is acceptable to use the fork to push food toward a sauce before eating it. Only small amounts of food should be eaten at a time

- Servers should watch for guests gesturing during conversation to prevent spillage and burns. Servers should indicate that service is coming to bring awareness to guests

- When the menu is preset, tables will be set with all the flatware needed. Many people are embarrassed to ask what to use first. As a rule, the first utensil used is the one placed at the outer right or outer left and then the order of use is inward so that by the time dessert is served, only the dessert spoon and fork (sometimes located at the top of the table setting) remains.

- When appetizers are served, their utensil may be brought with them and set on the plate in front of the guest.

- Servers can let confused guests know which utensil to use first. The teaspoon may be set out of sequence on the outside of the soup spoon.

- When a guest is finished eating the knife and fork can be placed to one side of the plate, either pointing in the same direction or crossed in the middle of the plate. It is wrong to place used flatware on the table or for servers to expect guests to reuse flatware. Fresh clean items should be preset or brought with each course.

- What should a server do if a guest uses the "wrong" item or seems to want to return a utensil that they will need for a future course? (Server should simply replace used utensil. Guest should not be embarrassed. It is possible an item was dropped, dirty or defective and the guest wants to have it replaced.)

- When drinking from a glass, the fingers should not be allowed to touch the rim. Goblets and stemmed glassware should be handled by the stem.

C. Etiquette for Specific Foods

- Many foods require specific treatment to eat properly; utensils, dishes and other items may be required. Servers should know the specific needs of these foods so people can eat them properly.

- If a guest does not know how to eat a food, it is not wrong to ask. Alternatively, a guest may wait and see what others do and follow their lead.

1. Finger Foods

   - Receptions, cocktail parties and other social events often feature situations where guests eat and drink standing up
   - Tables and chairs are optimal but small stands to set beverages while eating or bus stands where guests can place their empty items are desirable.
   - When left without options, guests may leave soiled dishes and glassware on any available surface
   - Why would there be a social event with food and drink but without tables and chairs? (Guests who are seated eat and drink more and take up more room, some of these functions are designed with mingling in mind and keeping costs down)
   - Moist or greasy foods soil guests' fingers so plenty of napkins, finger bowls or small cloths dipped in hot water and wrung dry should be provided.
   - Food served very hot should come with a warning to guests
   - Five inch bread and butter plates with/out a doily are often used at cocktail and receptions. Guests may serve themselves from a buffet or food items may be passed. Flying platters were carried over the heads of guests by Russians in the 19th century.
   - Guests should not double dip and avoid spilling

- Picnic food such as hot dogs, tacos and sub sandwiches should be wrapped in a napkin to hold it together and avoid spillage. The same holds true for many frozen desserts
- Beef and pork ribs can and should be eaten with the fingers. Extra napkins or moist towels should be provided. Other BBQ foods may receive the same service. Finger bowls should be served if used by the operation.
- It is proper to lift the skewer of shish ka bobs by the handle and push the pieces of food onto the plate with a fork
- How can the party planner assist guests at receptions besides providing a place to sit? (Various- menu might be mindful of ability to eat easily and consider mess factor.)

2. Soups

- Soup is served in a bowl or cup. If the cup has two handles, the guest can lift the cup when only a small amount of thin soup is left.
- Drinking from a bowl or blowing on soup to cool it is considered uncouth.
- Croutons should be served with a spoon, never the hands. It is better to eat crackers separately but they may be broken up and scattered over soup.
- Soup spoons are larger than other spoons. Move it across the bowl and away from self.
- The bowl can be tipped toward the far side where last of soup can be spooned up. When finished, the spoon is left on the bowl, not on the table.

3. Breads

- Loaves should be served on a cutting board with serrated knife and napkin with which to hold the loaf while cutting

4. Vegetables

- Corn on the cob can be held with one hand, both hands or with holders
- Artichokes are eaten by pulling off the leaves one at a time and dipping them into sauce, and then removing the soft flesh with the teeth. Leaves are placed on provided plate and the heart is eaten with a knife and fork after removing the thistles.
- Asparagus can be eaten with the fingers but a knife and fork are more proper

- A vegetable may be served with the sauce on the plate or the sauce may be brought in a sauce boat. The server should either serve guests their desired amount of sauce or allow them to serve themselves.

- Ask class members if they have eaten edamame and how they did so. (Steamed soy beans in the pod are served as an appetizer at sushi restaurants. It is picked up with the fingers where the salted pod is easily pressed open and the beans popped into the mouth. Used pods are placed on a plate for removal. Tempting as it is, salted fingers should not be licked)

5. Salads

- A knife may be used to cut large leaves, vegetables and meats.
- Sometimes a salad is served as a first course. If guests prefer to eat it with their main meal, the server should move the salad to the guest's left, the same is true if the guest wishes to finish their salad with their entrée
- If guests want dressing on the side, it should be served in a small container and placed to the side of the salad or on a small plate.

6. Pasta

- Some prefer to eat long pastas by twisting a small amount around the fork and supporting it with a large spoon. Or the fork can be placed almost vertically onto the plate and twisted
- Taking a small amount and bringing it to the mouth without twisting is preferred by some but it might be more difficult than the rolling method
- Is it possible these challenges amount, in part, for the popularity of various shaped pastas? (Various speculation)

7. Fish and Shellfish

- Some fish must be boned; they have a central bone structure to which most other bones are attached. The fins also carry bones. If not removed by filleting these must be removed after service.
- Fin bones are easily removed by placing a knife or fork under the flesh and removing it. By placing a knife or fork along the back and lifting, the flesh on top the central bone structure can be exposed and the bone lifted out.
- In fine dining a fish fork and knife are provided as are lemon wedges or halves. Care should be taken not to squirt juice or seeds.
- Oysters and clams are often served raw on the half shell, often loosened. They come on a plate with an oyster fork, cocktail sauce, lemon or warm butter. Some like hot sauce
- It is improper to cut clams or oysters, they should be eaten whole
- Steamed clams and mussels are often served in a large bowl along with oyster fork, sauce, warm lemon butter, oyster crackers and steaming liquor.

- The shell can be picked up and removed or left in the bowl and removed from there. An empty bowl should be provided for empty shells. A finger bowl is a customary service.
- Removing the meat is the most difficult part of eating crustaceans such as lobster or crab.
- To eat whole or half lobster, the claw should be cracked, before or after service and a cocktail fork or pick inserted to gently pull meat out
- Warm drawn or lemon butter is usually served.
- An extra plate should be provided to the right of the guest for empty shells
- A lobster bib is appropriate for lobster or any seafood item that must be shelled and a finger bowl should be provided after eating
- Very large cocktail shrimp present a challenge since they should not be partially eaten and put back. They should be cut with a fork rather than the teeth. Likewise with sushi or ceviche

8. Desserts

- In fine dining, the dessert fork and spoon are set together above the cover pointing in different directions. In many other operations, they are brought with the dessert
- Desserts that can be eaten with either a fork or spoon should have both brought so the guest can select which utensil to use
- Fruit is sometimes served for dessert and will have utensils appropriate to that item. A plate should be available for peels or cores
- Eating a piece of whole fruit by holding is called eating hand fruit
- At the end of the meal, the attention of the server is caught and the check is requested
- Why must the guest ask for the check at dinner? (It is considered rude to rush the guests out by offering it before asked)

IV. Tipping

- Tips should be based on the total before tax

- 10% is the least one should tip for adequate service. Below 10% indicates displeasure with the service

- 15% is normal for satisfactory service though some tip more

- 20% or higher may be given for unusually good service

- If servers are to split a tip, the tipper should indicate the division. 15-20% of the nontax total is often given to the captain

- People from outside the US are not used to tipping
- What do other countries do? (The gratuity is automatically added to the bill, much like any other sales commission)
- Valet parking and maitre'ds are often tipped
- An attentive server should not have to be asked for the bill
- Doesn't that contradict the previous statement? (The server should pay attention for signs that the guest is ready for the bill and present it upon signal)

V. Dining Etiquette of Various Cultures

- Rules of dining etiquette are not universal or consistent, even within a culture
- Ethnic dining has become an integral part of the foodservice industry. The most common cuisines are; Italian, Chinese, Mexican and more establishments are offering Indian, Japanese and Middle Eastern.
- What other ethnicities are represented in your geographic region? (Various)

A. Traditional Jewish Service

- Jewish dietary practices are interwoven with their faith
- Kosher means fit or proper and refers to dietary rules called Kashruth, biblically based laws applied by rabbinic interpretation
- Mammals must have split hooves and chew their food
- Both birds and mammals must be slaughtered in a specific manner called shechita by a trained slaughterer called a shotchet
- Meat must be Koshered or soaked and salted in a specific manner to remove all blood
- Only fish with both fins and scales are kosher, excluding shellfish
- Dairy products and meats cannot be cooked or served together and there are separate cooking utensils and serving dishes for meat and dairy
- No cooking is permitted on the Sabbath so food must be eaten cold or prepared ahead and kept warm

- Could any licensed caterer perform a Kosher catering function if they study these rules? (Not likely- must strictly follow the rules)

B. European Dining Etiquette

- For the most part, European etiquette is similar to the US. Traditionally it was more formal but even Europe is becoming more fast paced and casual

C. Chinese and Japanese

- Although there are many differences, both emerged thousands of years ago in common.

- Chopsticks are used in both

- Rice, wheat, soy sauce, and tofu are common in both

- The separation of these cultures for so many years have produced many differences

1. Chinese Dining

    - Chinese dining is adapted from family dining where food is brought to the table in serving dishes and guests help themselves.
    - Main plate is about the size of a salad plate and portions are small. There are also rice and a soup bowls. Rice wine may be served in small cups
    - There is typically no other liquid on the table because the soup supplies liquid though tea is sometimes served
    - Chopsticks are the main eating utensil as well as a special soup spoon
    - It is not considered impolite to raise the rice bowl to the mouth to guide the rice in with the chopsticks
    - The host traditionally sits with his back to the door with the main or honored guest opposite
    - It is considered an insult to the main guest to heave the grain of wood running toward him, it must run toward the other guests
    - Do students have an opinion about how Chinese service in the United States has often gone to the buffet? (Various)

2. Japanese Dining

    - The Japanese take great care in the presentation of their food with precise and decorative placement of food on the plate

- Glass dishes are often used in the summer, green porcelain in the spring, heavy stoneware in the winter and basket shaped dishes tinted with autumn leaves colors in the fall
- Traditional Japanese dining is quite formal
- The meal is usually served all at once
- It is considered bad taste to have any cutting utensils on the table. Soups and beverages are sipped from their dishes
- Japanese kneel on a tatami or mat and the food is served on a mall raised stand or a large low table for a group, The host faces the guests and no one is seated on the ends
- A wall scroll is a place of honor that the guest must wait for the host to insist on their being seated in front of
- Many restaurants provide small ceramic holders for chopsticks or they come wrapped in paper where the chopsticks can rest.
- Pickles are typically served at the end of the meal
- Which of these customs are typically practiced at sushi restaurants? (Chopsticks, beautiful presentation, soup, etc. various) Where does the Japanese steak house with the common grill fit in? (Various)

D. Indian Dining

- Indians are largely vegetarians. Some use no eating utensils but their right hands. Food is usually ladled out of bowls onto plates

- Many people sit on the floor to eat and no one wears shoes in the dining room

- Upper classes use utensils for dining

- Rice is eaten instead of potatoes. Pork is avoided and most of Hindu faiths do not eat beef.

- Ask students about their experience with Indian restaurants. (Various, also subject to buffet service)

E. Middle Eastern Dining

- Most Muslim diners follow traditional religious influences, food resources and social customs.

- It is customary to wash the hands before and after eating. One eats only with the right hand and often no utensils are used.

- As many as thirty different dishes are used in small quantities with cold food coming first

- Guests serve themselves, commonly using pita bread to help lift the food to the mouth

- When meat is eaten, it is usually followed with rice

- Guests are honored by being seated to the right of the host

- Ask students about their experience with Middle Eastern dining. (Various and less likely in more rural settings except for sandwich/casual service type operations)

Chapter Review Questions Answer Key

1. It is important for a server to know proper table manners to help the guest feel more comfortable by providing the food and beverages placed properly with the appropriate items for their consumption and to assist the guest who is puzzled as to how to proceed in a mannerly fashion.
2. The rules for proper eating developed grew out of what is practical and makes for good social functioning through the ideas of people themselves, codified by authors. They change over time as customs and cultures change and are not inviolate as they depend on the situation at hand.
3. Broiled lobster should be served with a lobster fork, lemons, butter sauce, a bib, plenty of napkins and finger bowls. A cracker might be used for cracking open the claws.
4. A finger bowl is a (monkey) dish of warm water with a lemon floating in it or a cloth dipped in hot water and wrung out to be used with moist or greasy finger foods.
5. When a guest is finished eating, the knife and fork can be placed to one side of the plate, both pointing in the same direction. Or with the two utensils crossed in the middle of the plate. Used flatware should not be placed on the table or be expected by servers to be reused.
6. Rules of etiquette may be violated when it makes sense to do so. It is always impolite to point out violations or to make violators feel uncomfortable. Many foods are finger foods such as sandwiches, corn on the cob, ribs, fried chicken and shell fish and they are properly picked up with the fingers in order to be eaten.
7. Some people prefer to eat long pastas by twisting a small amount of strings around the fork and supporting it with a large spoon. This is a customary and proper way to eat pasta.
8. A tip below 10% indicates displeasure with the service. A 10% tip is the least one should tip for adequate service. 15% is the normal tip and 20% or higher for unusually good service. Tips are based on the total before taxes.
9. A guest spill should be handled by the server first by avoiding its occurrence. This can be accomplished by warning guests when the server is near with dishes and to wrap picnic food such as hot dogs, burgers, subs and tacos in a napkin. If a spill does occur, the server should tidy up and cover the spill with a napkin.

10. One of the first things the server should ask the guests for is their drink order. If the guest does not care for an alcoholic beverage, the server should suggest a nonalcoholic beverage.

11. a. Finger bowls are for the purpose of tidying up the fingers when greasy or moist foods have been eaten
    b. Sharing is perfectly alright (unless against restaurant policy, such as on all you can eat salad bars when only one guest is paying). The server should have the kitchen split the order there to avoid spills or frustrations at the table.
    c. It is normally rude to start before everyone is served but banquets are an exception as the food of the first served would be cold by the time the last received theirs. Guests who have not yet been served should politely urge others to go ahead. It is proper to begin when served in this situation.
    d. Forks are set to be used from the outside in. Salad forks are generally the far left fork away from the plate
    e. Snails are served in a special indented dish resting in drawn butter, either in their shells or out. A snail fork is served with snails and often escargot tongs for grasping the shell while removing the meat. Lemon is typically served with the dish.
    f. Mussel and other shells should be placed in bowl or dish provided for that purpose by the server on the table to the right of the guest
    g. Yes, it is proper to use the fingers to eat ribs and other finger foods.
    h. To get the last of the soup, it is not improper to tip the bowl and let the soup run to the far side where it can be spooned up.
    i. Some people prefer to eat long pastas by twisting a small amount of strings around the fork and supporting it with a large spoon or the fork can be placed down almost vertically on the plate and twisted until the desired amount is obtained.
    j. Oysters are often loosened from the shell for the guest's convenience. They usually come on the plate set in a circle, resting on a bed of ice with an oyster fork. Cocktail sauce, lemon or warm butter may be served with them. Some guests like hot sauce, in addition. It is improper to cut up clams or oysters. They should be eaten whole. They can be dipped into sauce or have sauced dropped on them. It is acceptable to crush crackers into the sauce.

Chapter Cases Answer Key

Serving Fish and Seafood

- Fish have a central bone structure to which other bones are attached. The fins may also carry bones. To remove these, place a knife of fork under the flesh surrounding the fin and remove it. Place a fork or knife along the back and lift, exposing the central bone structure and lifting this bone out. This may done to relieve the guest of this process, tableside. A fish fork and knife should be provided. Lemon wedges or halves should be served wrapped in cheesecloth to prevent the seeds and juice from squirting.
- Oysters and clams are often served raw, loosened from the shell. They come with cocktail sauce, and an oyster fork, lemon or warm butter. Hot sauce may be desired. Oyster crackers may be served also.

- Steamed clams and mussels are often served in a large bowl with oyster fork, sauce, warm lemon butter, oyster crackers and stewing liquor. An empty bowl for shells and finger bowl should be brought for the guests.
- Lobster and crab present difficulties to guests regarding removing the meat. Remove the meat for the guests, first cracking the shell, or provide the guest with a cracking tool and cocktail fork. These crustaceans are served with warm butter and lemon and an extra plate or bowl to the right of the guest. A lobster bib should be offered as well as a finger bowl. Shrimp cocktail should be served with a cocktail fork. Large pieces of shrimp or sushi may require a knife.

The Tipping Problem

The person who left a 10% tip left the least possible amount for merely adequate service. Excellent service should receive a minimum of 15% with many guests leaving 20% or more. As servers often receive the bulk of their remuneration in the form of gratuities, the person who left the inadequate tip is, in effect, cheating Mary out of earned income. (It is doubtful that the person leaving the poor tip would be satisfied with receiving their own income in such a manner).

Chapter 11 Test

True/False

T F 1. The first written record of rules of decorum dates back to the time of written history thousands of years before Christ.

T F 2. Rules of etiquette are set standards that are not subject to change.

T F 3. Today's rules of etiquette are less formal than they were during England's Victorian era.

T F 4. The zig zag method is a European custom of cutting meat on the plate with a sharp serrated knife.

T F 5. It is never proper to use the fingers to eat.

Multiple Choice

6. "To possess good manners is to not tell others what good manners are" means, in terms of table service:
   a. The "snooty" server that looks down their nose at the guest is ill-mannered
   b. Guests must know what good manners are and practice them without needing cues from the server
   c. Guests without good manners should not expect to be treated respectfully by members of their party
   d. Guests without good manners should not expect to be treated respectfully by members of the restaurant service staff

7. The first known written record of rules of decorum was thought to be:
   a. "A Treatise on Manners"
   b. "The Courtier"
   c. "Health and enjoyment within decency"
   d. "Miss Manners Guide to Excruciatingly Correct Behavior"

8. Which of the following describes the authors of etiquette rule books?
   a. They were uncommonly well mannered and wished to share their gifts with the world of us stumbling bumpkins.
   b. They wore large feathered hats and were squeamish about the behavior of others at the table
   c. They created rules to dine by due to their superior social status and worldly experience
   d. They collected, summarized and published rules formed by people

9. Which of the following is NOT proper napkin etiquette?
   a. The guest tucking the napkin under their chin
   b. The server placing napkins into the guests laps for them
   c. The guest discreetly fastening the napkin into their belt so it will stay put

d. The guest taking the gently unfolded napkin from the edge of the table where it was put by the server and placing it on their lap

10. When a guest places flatware on the soiled plate from a freshly consumed course and lacks the utensils to eat the next course, the server should:
    a. Place the flatware from the plate back on the table to use in the course to come.
    b. Take away the plate and flatware and bring the next course. The guest will be left without flatware and will learn not to give back needed utensils. Of course the server will bring new utensils when the guest requests replacements.
    c. Remove the plate and flatware and bring replacements prior to serving the next course or simultaneously.
    d. Assume that the guest is full and pack up the next course "to-go".

11. Which of the following statements is false, regarding proper seafood eating etiquette?
    a. It is proper to wear a bib to eat lobster and other messy shellfish items that require cracking
    b. It is proper to cut a large raw oyster into two pieces
    c. It is alright for the guest to remove a fish bone from their mouth using their fingers
    d. It is proper to cut a large shrimp cocktail shrimp into pieces to eat it though only with a fork, not with the teeth.

12. Which of the following is the customary gratuity for satisfactory service?
    a. Gratuities are optional and not necessary for service that is merely satisfactory.
    b. 10%
    c. 15% or more
    d. A dollar per person is always an acceptable amount to tip

13. Which of the following statements is false regarding traditional Jewish Service?
    a. Mammals must have split hooves and chew their cuds
    b. Meat and dairy products cannot be cooked or served together
    c. Fish need to be killed in specific manner by a trained Jewish slaughterer
    d. Meat must be soaked and salted in a specific manner to remove all blood

14. All of the following are commonalities between Japanese and Chinese service EXCEPT:
    a. Guests use chopsticks to eat
    b. Guests typically kneel on a mat and the food is served on low tables
    c. Rice, wheat, soy sauce and tofu are typically found
    d. Guests are honored by their particular place at the table

15. All of the following characterizes Middle Eastern Dining EXCEPT:
    a. Guests are honored by being seated to the right of the host
    b. One eats only with the right hand and often no utensils are used
    c. Pita bread is commonly used to help lift the food to the mouth
    d. Guests are served by women dressed in traditional costumes

Short Answer/Essay

16. How can a guest courteously gain the attention of the server?
17. Describe the etiquette issues created when the server does not bring an entire tables food out at the same time.
18. Draw the proper place setting for a banquet diner being served the following: Soup, Salad, Entrée featuring boneless chicken and a dessert pastry. Wine will be served with the entrée as well as dinner rolls. Place all utensils in your rendering in their proper places.
19. What does tipping have to do with etiquette?
20. Are rules of etiquette Universal? Explain.

Chapter 11 Test Answer Key

True/False

1. False. The first record of written rules of decorum is dated 1474: "Health and Enjoyment within Decency."
2. False. Rules of etiquette change constantly as society changes.
3. True
4. False. The zig zag method refers to the US custom of holding the knife with the right hand and fork with the left and then transferring the fork to the right hand to bring the food to the mouth. Europeans use their left fork holding hands to bring food to the mouth.
5. False. Oysters on the half shell, fried chicken and ribs can and should be eaten with the fingers. Other foods such as sandwiches are also designed to be eaten with the hands.

Multiple Choice

6. a
7. c
8. d
9. a.
10. c
11. b
12. c
13. c
14. b
15. d

Short Answer/Essay

16. Guests should raise one hand or call "server" in a soft voice to get the attention of their waitperson. It is considered rude to whistle, clap, snap fingers, strike a glass or cup, stamp feet or make other noises to attract a server
17. The guests who have been served feel it is impolite to begin eating until everyone has their food. If they wait, their food may well be cold or less desirable. Those who have not been served must urge the guests who have been to begin, while not having food themselves. The table is then unlikely to finish this course together, leaving those who are done with dirty plates in front of them. It is considered rude to clear until everyone is finished. Unless it is a very large party or banquet, the server should everyone at a table at the same time.
18. B&B plate with knife (handle facing guest, blade down), Salad fork, Dinner fork, Napkin, Dinner knife; blade facing in, Soup Spoon, and teaspoon and dessert fork above the setting. A wine glass would sit to the upper right of the dinner knife. (Variations mentioned in the text would also be accepted)
19. Tipping is the customary method of a guest expressing appreciation to the server. The amount of tip varies with the service provided. Below 10% indicates displeasure with the server, 10% represents only adequate service and is considered the least amount to be left. 15% is the normal amount of satisfactory service, though that may be more. 20% or more

may be left for unusually good service. Servers often split tips according to function. Valet parking and Maitre d's can be tipped. People outside the United States are not used to tipping. Tipping is included in an etiquette chapter as many people do not know what percentages are customary. Some believe that tips are entirely discretionary, perhaps because they are unaware how much of the income of many servers are derived from this source.

20. Rules of dining etiquette are not universal. Cultures have vastly different dining rules from cultures that use no utensils to those that use many. Dining rules are not always consistent within a culture. Dietary practices may be interwoven with faith. Family dining, the honoring of guests and seated on the floor customs are all variations on what is familiar to those in the US.

Activity/Training

Have class members' list things that are their "pet peeves" when it comes to the public dining behavior of others. Have them match each item on their list with something a server could do to help eliminate the problem.

Alternatively, have class members write up a handbook of etiquette practices that relate to your lab, if there is one and the menu items served. Class members could go further and conduct business and social dining etiquette presentations for class members or members of another related class.

Made in United States
Orlando, FL
04 November 2026

90200454R00114